# THE CANCER CHRONICLES

ALSO BY GEORGE JOHNSON

*The Ten Most Beautiful Experiments*

*Miss Leavitt's Stars: The Untold Story of the
Woman Who Discovered How to Measure the Universe*

*A Shortcut Through Time: The Path to the Quantum Computer*

*Strange Beauty: Murray Gell-Mann and the Revolution
in Twentieth-Century Physics*

*Fire in the Mind: Science, Faith, and the Search for Order*

*In the Palaces of Memory: How We Build the Worlds Inside Our Heads*

*Machinery of the Mind: Inside the New Science of Artificial Intelligence*

*Architects of Fear: Conspiracy Theories and Paranoia in American Politics*

# The Cancer Chronicles

Unlocking Medicine's Deepest Mystery

George Johnson

THE BODLEY HEAD
LONDON

Published by The Bodley Head 2013

2 4 6 8 10 9 7 5 3 1

Copyright © George Johnson 2013

George Johnson has asserted his right under the Copyright, Designs
and Patents Act 1988 to be identified as the author of this work

First published in Great Britain in 2013 by
The Bodley Head
Random House, 20 Vauxhall Bridge Road,
London SW1V 2SA

www.bodleyhead.co.uk
www.vintage-books.co.uk

Addresses for companies within The Random House Group Limited can
be found at: www.randomhouse.co.uk/offices.htm

The Random House Group Limited Reg. No. 954009

A CIP catalogue record for this book
is available from the British Library

ISBN 9781847921666 (Hardback)
ISBN 9781847921673 (Trade paperback)

The Random House Group Limited supports the Forest Stewardship Council® (FSC®), the leading
international forest-certification organisation. Our books carrying the FSC label are printed on
FSC®-certified paper. FSC is the only forest-certification scheme supported by the leading
environmental organisations, including Greenpeace. Our paper procurement policy
can be found at: www.randomhouse.co.uk/environment

Printed and bound in Great Britain by
Clays Ltd, St Ives PLC

For Joe's girls,

Jennifer, Joanna, Jessica, and Emmy

and for his wife, Mary Ann

---

*We must never feel disarmed: nature is immense and complex, but it is not impermeable to intelligence; we must circle around it, pierce and probe it, looking for the opening or making it.*

—PRIMO LEVI, *The Periodic Table*

# Contents

# Author's Note

Several years ago, for reasons that will become clear in these pages, I was driven to learn everything I could about the science of cancer. How much could I as an outsider, a longtime science writer more comfortable with the sharp edges of cosmology and physics, grasp of this wet, amorphous, and ever-changing terrain? I imagined the expanse before me as a boundless rain forest whose breadth and diversity could never be captured within a single book or even a single mind. I would find an opening at one of the borders and enter, cutting my own path, exploring where my curiosity led—until I emerged years later at the other side, with a better understanding of what we know and don't know about cancer. I was in for some remarkable surprises.

Many people helped along the way. First I thank the scientists who devoted so much time—sitting for interviews, answering e-mails, reviewing parts or all of the manuscript: David Agus, Arthur Aufderheide, Robert Austin, John Baron, José Baselga, Ron Blakey, Timothy Bromage, Dan Chure, Tom Curran, Paul Davies, Amanda Nickles Fader, William Field, Andy Futreal, Rebecca Goldin, Anne Grauer, Mel Greaves, Seymour Grufferman, Brian Henderson, Richard Hill, Daniel Hillis, Elizabeth Jacobs, Scott Kern, Robert Kruszinsky, Mitchell Lazar, Jay Lubin, David Lyden, Franziska Michor, Jeremy Nicholson, Elio Riboli, Kenneth Rothman, Bruce Rothschild, Chris Stringer, Bert Vogelstein, Robert Weinberg, Tim White, and Michael Zimmerman. In addition I consulted more than five hundred papers and books about cancer and sat in on dozens of lectures. Most of these sources are listed as references in my endnotes along with interesting information that didn't make it into the main text. George Demetri and Margaret Foti kindly allowed me to sit in on a private workshop in Boston organized by the American Association for Cancer Research. Thanks to them and the staff of AACR, including Mark Mendenhall and Jeremy Moore, who welcomed me to the organization's fascinating annual meeting in Florida. I am also grateful to the Keystone Symposia and the Society for Developmental Biology for accommodating me at some of their events.

Just as I was getting my boots wet, David Corcoran at *The New York Times* enthusiastically commissioned and published two of my early reports. Thanks to him and other colleagues—Christie Aschwanden, Siri Carpenter, Jennie Dusheck, Jeanne Erdmann, Dan Fagin, Louisa Gilder, Amy Harmon, Erika Check Hayden, Kendall Powell, Julie Rehmeyer, Lara Santoro, Gary Taubes, and Margaret Wertheim—for their reactions and advice on the manuscript.

Several recent alumni of the Santa Fe Science Writing Workshop read early versions, offering their good sense and expertise: April Gocha, Cristina Russo, Natalie Webb, Shannon Weiman, and Celerino Abad-Zapatero. Bonnie Lee La Madeleine and Mara Vatz helped with library research and the endless checking of facts. The manuscript was in constant flux and any errors that survive are my own. This will be the seventh book I have done with Jon Segal, my editor at Knopf, and the fourth with Will Sulkin of Jonathan Cape and Bodley Head in London. Thanks to them and their colleagues—including Victoria Pearson, Joey McGarvey, Meghan Houser, and Amy Ryan, a superb copyeditor—and to Esther Newberg, my agent almost from the start.

Special thanks to Cormac McCarthy, who read an early version of the book, and to Jessica Reed, whose literary sensibility and encouragement were an inspiration. More than once my friend Lisa Chong read through the book sentence by sentence, page by page, helping to apply the finishing touch.

Finally my deep thanks to Nancy Maret and the family of my brother, Joe Johnson, who allowed me to tell their stories.

# The Cancer Chronicles

*I wonder now, though, if the steady presence of music around me didn't contribute importantly to my sense of the cancer as a thing with its own rights. Now it sounds a little cracked to describe, but then I often felt that the tumor was as much a part of me as my liver or lungs and could call for its needs of space and food. I only hoped that it wouldn't need all of me.*

—Reynolds Price, *A Whole New Life*

*Tuberculosis used to be called "consumption" because it consumes. It dissolved a lung or bone. But cancer produces. It is a monster of productivity.*

—John Gunther, *Death Be Not Proud*

Chapter 1

# Jurassic Cancer

As I crossed a dry, lonesome stretch of the Dinosaur Diamond Prehistoric Highway, I tried to picture what western Colorado—a wilderness of sage-covered mesas and rocky canyons—looked like 150 million years ago, in Late Jurassic time. North America was breaking away from Europe and Asia—all three had formed a primordial supercontinent called Laurasia. The huge land mass, flatter than it is today, was drifting northward a few centimeters per year and was passing like a ship through the waters of what geographers would come to call the Tropic of Cancer. Mile-high Denver was near sea level and lay about as far south as where the Bahamas are today. Though the climate was fairly dry, webs of rivulets connecting shallow lakes and swamps covered part of the land, and vegetation abounded. There were no grasses or flowers—they had yet to evolve—just a weird mix of conifers commingling with ginkgos, tree ferns, cycads, and horsetails. Giant termite nests soared as much as thirty feet high. Splashing and stomping through this Seuss-like world were *Stegosaurus, Allosaurus, Brachiosaurus, Barosaurus, Seismosaurus*—their bones buried far below me as I made my way from Grand Junction to a town called Dinosaur.

Occasionally one can glimpse outcroppings of the Jurassic past, exposed by erosion, seismological uplift, or a highway department road cut—colorful bands of sediment that form a paleontological treasure house called the Morrison Formation. I knew what to look for from photographs: crumbling layers of reddish, grayish, purplish, sometimes greenish sediment—geological debris piled up over some 7 million years.

Just south of the town of Fruita on the Colorado River, I hiked to the top of Dinosaur Hill, stopping for a moment to pick up a pinch of purplish Morrison mudstone that had fallen near the trail. As I rolled it in my fingers it crumbled like dry cookie dough. On the far side of the hill, I came to a shaft where in 1901 a paleontologist named Elmer Riggs extracted 6 tons of bones that had belonged to an *Apatosaurus* (the proper name for what most of us call a *Brontosaurus*). Alive and fully hydrated, the 70-foot-long reptile would have weighed 30 tons. Riggs encased the bones in plaster of paris for protection, ferried them across the Colorado on a flat-bottom boat, and then shipped them by train to the Field Museum in Chicago, where they were reassembled and put on display.

After making my way north to Dinosaur (population 339), where Brontosaurus Boulevard intersects Stegosaurus Freeway, I stood at an overlook and watched Morrison stripes in a canyon reddening with the setting sun. But it was a little farther west, along the Green River in the western reaches of Dinosaur National Monument, that I saw the most beautiful example: a cliffside of greenish grays slumping into purples slumping into browns. It indeed resembled, as the woman at the park headquarters had told me, melted Neapolitan ice cream.

It was somewhere in these parts that a dinosaur bone was discovered that displays what may be the oldest known case of cancer. After the dinosaur died, whether from the tumor or something else, its organs were eaten by predators or rapidly decomposed. But the skeleton—at least a piece of it—gradually became buried by wind-blown dirt and sand. Later on, an expanding lake or a meandering stream flowed over the debris, and the stage was set for fossilization. Molecule by molecule minerals in the bones were slowly replaced by minerals dissolved from the water. Tiny cavities were filled and petrified. Several epochs later dinosaurs were long extinct, their world overlaid by lakes and deserts and oceans, but this fossilized bone, encased in sedimentary rock, was preserved and carried through time.

That hardly ever happened. Most bones disintegrated before they could become fossilized. And of the fraction that survived long enough to petrify, all but a few remain buried. The specimen, now labeled CM 72656 and housed at the Carnegie Museum of Natural History in Pittsburgh, was a survivor. Unearthed by a rushing river or exposed by tectonic forces—somehow it was delivered to the surface of our world where, 150 million years after the animal died, it was discovered by some forgotten rockhound. A cross-section was cut with a rock saw, polished, and after passing through who knows how many human hands, the fossil found its way to a Colorado rock shop where it caught the eye of a doctor who thought he knew a case of bone cancer when he saw one.

His name was Raymond G. Bunge, a professor of urology at the University of Iowa College of Medicine. In the early 1990s, he telephoned the school's geology department to ask if someone would come evaluate a few prize specimens in his collection. The call made its way through the switchboard to Brian Witzke, who on a cold autumn day bicycled to the doctor's house and was presented with an attractive chunk, 5 inches thick, of mineralized dinosaur bone. Viewed head-on, the fossil measured 6.5 by 9.5 inches. Lodged inside its core was an intrusion, now crystallized, that had grown so large it had encroached into the outer bone. Bunge suspected osteosarcoma—he had seen the damage the cancer can do to human skeletons, particularly those of children. Oval in shape and the size of a slightly squashed softball, the tumor had been converted over the millennia into agate.

The fragment was too small for Witzke to identify the bone type or the species of dinosaur, but he was able to provide a geological diagnosis: The reddish-brown color and the agatized center were clues that it came from the Morrison Formation. Bunge remembered buying the souvenir somewhere in western Colorado—burnished pieces of petrified dinosaur bone were a favorite among collectors—but he couldn't remember the precise location. He gave the rock to the geologist, asking that he seek an expert opinion.

Other projects intervened, and so the fossil sat almost forgotten atop a filing cabinet in Witzke's office, until the day he sent it to Bruce Rothschild, a rheumatologist at the Arthritis Center of Northeast Ohio who had expanded his practice to include dinosaur bone disease. He had never seen a clearer or more ancient example of prehistoric cancer. His next step was to determine just what kind of cancer it was.

The tumor, it turned out, didn't exhibit the ill-defined margins or the layered, onion-skin look of an osteosarcoma, the cancer Bunge had suspected, or of another malignancy called Ewing's sarcoma. Rothschild also felt confident in ruling out myeloma, a cancer of plasma cells that leaves bone with a "punched out" appearance. The fact that the tumor, gnawing its way outward, had left intact a thin shell of bone was reason to exclude the more invasive multiple myeloma. Every skeletal disease leaves a distinct engraving and, one by one, Rothschild eliminated the possibilities: "the superficial solitary and coalescing pits of leukaemia," "the expansile, soap bubble appearance of aneurysmal bone cysts," "the epiphyseal 'popcorn' calcifications characteristic of chondroblastomas," "the 'ground glass' appearance of fibrous dysplasia."

For an outsider reading Rothschild's observations, the medical jargon might be somewhere between translucent and opaque, words that gain a grim familiarity only as one strives to understand the sudden disruption of cancer. What is clear from the beginning is the confidence with which a specialist in the obscure discipline of dinosaur pathology can provide a likely diagnosis for a 150-million-year-old tumor. Rothschild went on to rule out the "sclerotic-rimmed lesions of gout," the "zones of resorption characteristic of tuberculosis," and the "sclerotic features of gummatous lesions of treponemal disease." Unicameral bone cysts, enchondromas, osteoblastomas, chondromyoxoid fibromas, osteoid osteoma, eosinophilic granuloma—who would have known that so much can go wrong inside what appears to be solid bone? None of these seemed like candidates. To Rothschild's eye the lesion had the markings of a metastatic cancer, the deadliest kind—a cancer that had originated from cells elsewhere in the dinosaur's body and migrated to establish a colony in the skeleton.

There had been scattered references in the journals to other dinosaur tumors—osteomas (clumps of overeager bone cells outgrowing their rightful bounds) and hemangiomas (abnormal effusions of blood vessels that can form within the spongy tissue inside bone). Like cancer, these benign tumors are a kind of neoplasm (from the Greek for "new growth")—cells that have learned to elude the body's checks and balances and exert a will of their own. The cells in a benign tumor are multiplying rather slowly and have not acquired the ability to invade surrounding tissue or to metastasize. They are not necessarily harmless. Occasionally a benign tumor can press dangerously against an organ or blood vessel or secrete destructive hormones. And some can become cancerous. These were rare enough. But sightings of malignant dinosaur tumors were especially scarce. A cauliflower-like growth in the forelimb of an *Allosaurus* was thought for a while to be a chondrosarcoma. But on close examination Rothschild decided that it was just a healed fracture that had become infected. Bunge's fossil was the real thing. In a terse, five-hundred-word paper written with Witzke and another colleague and published in *The Lancet* in 1999, he came to a bold conclusion: "This observation extends recognition of metastatic cancer origins to at least the mid-Mesozoic [the Age of the Dinosaurs], and is the oldest known example from the fossil record."

I'd first heard of Raymond Bunge's fossil earlier that summer when I began working my way through the literature on the science of cancer. There is something sickly fascinating about the way a single cell can break from the pack and start multiplying, creating something alien inside you—like a new organ suddenly sprouting in the wrong place or, even more gruesome, a vicious, misshapen embryo. Teratomas, rare tumors that arise from misguided germ cells (the ones that give rise to eggs and sperm), can contain the rudiments of hair, muscle, skin, teeth, and bone. Their name is from the Greek word *teras,* for "monster." A young Japanese woman had an ovarian cyst with head, torso, limbs, organs, and a cyclopean eye. But these cases are very rare. Tumors almost always evolve according to their own impromptu plan. The most dangerous ones become mobile. Once they have established themselves in the immediate vicinity— your stomach, your colon, your uterus—they move on, metastasize, to new ground. A cancer that began in the prostate gland can end up in the lungs or the spinal column. There was no reason to believe that cancer hadn't occurred in dinosaurs. But considering the tiny fraction of paleontological remains that humans have had the opportunity to examine, coming across an actual example seemed almost miraculous.

Consider the size of the field: From Dinosaur National Monument in Utah and Colorado, the Morrison Formation reaches north into Wyoming, Idaho, Montana, the Dakotas, and southern Canada. It spreads east to Nebraska and Kansas, and south to the panhandles of Texas and Oklahoma, and into New Mexico and Arizona. It covers approximately half a million square miles. Erosion and excavation, natural or man-made, have only nicked the edges, barely sampling the 7-million-year accumulation of dinosaur bones, and only those that happened to become fossilized. If it hadn't been for Raymond Bunge's sharp eye, the earliest solid evidence of prehistoric cancer would have been missed. How many other cases were crushed inside those lightless layers? And among the bones that have been retrieved, how many malignancies had been overlooked? Paleontologists were hardly ever looking for cancer—few would recognize it if they saw it—and the only tumors they had a chance of finding would be those that had tunneled their way outward to a bone's surface or had been revealed by a random fracture or the blind cut of a lapidary saw.

One of the most elusive questions about cancer is how much is timeless and inevitable—arising spontaneously inside the body— and how much has been brought on by pollution, industrial chemicals, and other devices of man. Getting a rough sense of the frequency of cancer in earlier epochs might provide important clues, but only with a larger sample of data. His interest piqued by Bunge's fossilized tumor, Rothschild began looking for more.

With a portable fluoroscope, he began x-raying his way through the museums of North America. In people, cancers that metastasize to the skeleton most commonly lodge in the spine, so Rothschild concentrated on vertebrae. By the time he was done he had examined 10,312 vertebrae from about seven hundred dinosaurs collected by the American Museum of Natural History in New York, the Carnegie Museum in Pittsburgh, the Field Museum in Chicago, and other institutions throughout the United States and Canada—every specimen north of the Mexican border that he could get his hands on. He inspected loose vertebrae and, using ladders and a cherry picker, the soaring spines of whole skeletons. (There is a picture of him wearing a dinosaur T-shirt and leaning backward inside the rib cage of a *Tyrannosaurus rex*.) Bones that appeared abnormal under x-rays were scrutinized more closely with a CT scan.

In the end, his diligence paid off. He found another bone metastasis, and this time it was possible to identify the victim: an *Edmontosaurus,* a duck-billed titan (the family name is Hadrosauridae) that lived toward the end of the Cretaceous, right after the Jurassic, when dinosaurs began to go extinct. Other Hadrosauridae also had bone tumors, all of them benign: an osteoblastoma, a desmoplastic fibroma, and twenty-six hemangiomas, but there were none among the other beasts. That perhaps was the biggest surprise. Although Hadrosauridae vertebrae made up less than one-third of the bone pile—about 2,800 specimens from fewer than one hundred dinosaurs—they were the source of all the tumors. The approximately 7,400 vertebrae that were not hadrosaurs—*Apatosaurus, Barosaurus, Allosaurus,* and so forth—exhibited no neoplasms, either malignant or benign.

It was the kind of anomaly epidemiologists of human cancer confront all the time. Why do some people get more cancer than others? Some evolutionary twist may have left *Hadrosaurus* with a genetic predisposition for tumors. Or the reason might have been metabolic. These dinosaurs, Rothschild speculated, may have been more warm-blooded than other ones. Warm-blooded metabolisms run faster—it takes energy to maintain body heat—and that might accelerate the accumulation of the cellular damage that leads to malignancy.

Maybe the difference was not endemic but environmental—something about what *Hadrosaurus* ate. Plants in an ecosystem engage in endless chemical warfare, synthesizing herbicides and insecticides to fight off pests. Some of these chemicals are mutagens: they can change DNA. Modern descendants of the fernlike cycads that grew in Mesozoic times produce poisons that can induce liver and kidney tumors in laboratory rats. But why would *Hadrosaurus* eat more cycads than, say, *Apatosaurus*? Another possible source of carcinogens—needles from conifer trees—had been discovered in the stomachs of a couple of *Edmontosaurus* "mummies," whose remains had been buried under the right environmental conditions to fossilize instead of rot. But that wasn't much evidence to go on.

There were other curiosities to explain. When *Hadrosaurus* tumors did occur it was only among the caudal vertebrae—those nearest the tail of the spine. What was it about the bottom of the reptile that was more susceptible than the top? If only dinosaurs could be re-created from ancient DNA as they were in *Jurassic Park* and made available for medical research. At the great cancer centers—Dana-Farber in Boston, MD Anderson in Houston, and others around the world—a scientist can consume a career studying the role a single molecule plays in malignancy. Just the data from Rothschild's survey suggested dissertations' worth of questions. The overriding one was how to put his findings into perspective. Human bone cancer of any kind—metastatic or originating in the skeleton—is a rarity. Was one case among seven hundred dinosaur skeletons a little or a lot?

In a third paper, Rothschild considered the odds. He had been approached by two astrophysicists who were hoping to support their theory that the end of the dinosaurs' earthly reign was hastened by a spike of radioactive cosmic rays. Ionizing radiation—the kind strong enough to damage DNA—can cause cancer, and bone marrow is particularly susceptible. If a cosmic event had unleashed unusually strong rays, the effect on the dinosaurs would have been like being x-rayed from outer space.

But how would you do the epidemiology? In an earlier study Rothschild and his wife, Christine, had x-rayed bones at the Hamann-Todd Human Osteological Collection at the Cleveland Museum of Natural History, a repository of three thousand skeletons from medical school cadavers—homeless souls who would otherwise be in pauper's graves. Thirty-three of them had metastatic bone tumors, which amounts to 1.14 percent. Autopsies at the San Diego Zoo suggested that reptiles have a bone cancer rate about one-eighth that of humans, or about 0.142 percent. One cancerous *Edmontosaurus* among seven hundred fluoroscoped dinosaurs yields almost precisely the same number. One would have to look elsewhere for evidence that cancer had been a factor in the extinction.

For months factoids like this had been accumulating in my notebook and metastasizing through my mind. Every question raised about cancer inevitably spawned more. How representative was the Hamann-Todd Collection of the overall cancer rate? The indigents whose bones were there may have suffered from poor nutrition and haphazard diets, possibly increasing their susceptibility. Yet many of them probably had comparatively short life spans, dying from violence or infectious diseases before there was time for a cancer to grow. Maybe it all balanced out. And maybe not. The study of the animals in the San Diego Zoo raised more questions. Animals in captivity tend to get more cancer than those in the wild, maybe because they are exposed to more pesticides or food additives, or maybe just because they survive longer, get less exercise, and eat more. Of all the risk factors associated with human cancer two that are seldom disputed are obesity and old age.

The most troubling question was how much one can extrapolate about dinosaur cancer—and the ultimate origins of the disease— from what little evidence has survived. If you included in the sample only the one hundred tumor-prone *hadrosaurs,* their bone cancer rate would be 1 percent, about the same as for the human skeletons. But you have to wonder how many other specimens are waiting to be discovered. Just one more with a malignant tumor would double the cancer rate. Finally there was the question of how many cancers might have spread to unexamined parts of the skeleton or to softer organs—cancers that never reached bone. Once the tissues decomposed the evidence would be gone.

There are reports of a possible exception. In 2003, the year the Rothschild survey appeared, paleontologists in South Dakota announced the discovery of what might be a dinosaur brain tumor. They were preparing the skull of a 72-million-year-old *Gorgosaurus,* a close relative of *Tyrannosaurus rex,* when they found "a weird mass of black material in the brain case." Analysis with x-rays and an electron microscope indicated that the rounded lump had consisted of bone cells, and veterinary pathologists diagnosed it as an "extraskeletal osteosarcoma," a bone-cell-producing tumor that had taken up residence in the cerebellum and brainstem. Maybe that explains why the *Gorgosaurus* appeared to be so battered, as though the animal, suffering from a loss of motor control, had stumbled and fallen repeatedly. "It certainly would take a bizarre event to have created this appearance," Rothschild speculated at the time. "The position and character may well be a tumor, but it still needs to be proven that this is not simply broken skull fragments that fell in."

Continuing along the Dinosaur Diamond Highway, thinking about cancer, I made my own rare sighting: a Sinclair gasoline station with its green dinosaur logo—another relic of earlier times. Along the road, rocking oil wells pumped the fossil fuels derived, as best we know, from prehistoric organic matter, a puree of tiny plant and animal life, perhaps with some oil of dinosaur splashed in.

It was almost dusk when I reached the Yampa Plateau in northern Colorado, a 300-million-year pile of geology. Eons of seismic turmoil—the thrusting and tilting, the slipping and sliding of great crustal masses—had made a mess of the timeline. For miles the road skimmed the surface of rock laid down in the Jurassic and Cretaceous, mid to late dinosaur time. Then without so much as a bump of the tires, the mesa top abruptly changed to Pennsylvanian—whole epochs sheared off to expose an older world, 150 million years before the Morrison dinosaurs, when primitive cockroaches crawled the land. Crushed a couple of strata beneath the Pennsylvanian would have been the Devonian, a 400-million-year-old countryside. In Devonian rock 1,600 miles east of the Yampa, a jawbone of a primitive armored fish was discovered near what became Cleveland, Ohio. It is pitted with what some scientists take to be a tumor and others dismiss as an old battle wound.

The road ended at Harpers Corner—the far tip of the plateau. I walked to the edge where deep below me the Green and Yampa Rivers come together, having sawed though all that hardened time. I stood there flummoxed by the thought of all that vanished past. After the disappearance of the dinosaurs came the Laramide orogeny, when the peaks that became the Rockies soared from the earth, reaching as high as 18,000 feet, only to become buried to their necks in their own debris. With the Exhumation of the Rockies (these names sound almost biblical), the infill began washing away. In early Pleistocene time, just 2 million years ago, the great glaciations followed, leaving behind the geography we know today. Throughout all of these cataclysms life kept evolving. Stowing away on the journey was this interloper called cancer.

Hints of benign neoplasms have been found in the fossilized bones of ancient elephants, mammoths, and horses. Hyperotosis, or runaway bone growth, appears in fish from the genus *Pachylebias,* which seem to have put the tumors to good use. With the ballast provided by the increased bone mass, the fish could graze deeper in the salty Mediterranean waters, giving them an edge over their competitors. What began as a pathological growth may have been adopted as an evolutionary strategy.

Malignant tumors have been suspected in an ancient buffalo and an ancient ibex. There is even a report from 1908 of cancer in the mummy of an ancient Egyptian baboon. The examples are scant and sometimes controversial. But as with the dinosaurs, absence of evidence is not evidence of absence. Maybe cancer was a great rarity before man began messing with the earth. But a core amount of cancer must have existed all along. For a body to live, its cells must be constantly dividing—splitting into two cells, which split into four, then eight, doubling again and again. With each division the long threads of DNA—the repository of a creature's genetic information—must be duplicated and passed along. Over the course of time mechanisms have evolved to repair errors. But in a world awash with entropy that is naturally an imperfect process. When it goes wrong the result is usually just a dead cell. But under the right circumstances the errors give rise to cancer.

Even a lone single-celled bacterium can spawn a mutation that causes it to replicate more vigorously than its neighbors. When that happens to a cell within a tissue the result is a neoplasm. Plants and animals—two variations on the theme of multicellularity—ultimately sprang from the same primordial source. Plants are our very distant cousins, and they do get something resembling cancer. A bacterium called *Agrobacterium tumefaciens* can transfer a fragment of its own DNA into the genome of a plant cell, causing it to multiply into a tumor called crown gall. A remarkable paper published in 1942 demonstrates that in sunflowers these tumors can spawn secondary tumors—a primitive analog of metastasis. In the insect world larval cells can give rise to invasive tumors—the same phenomenon, perhaps, that carried over to the vertebrates.

Cancer (sarcomas, carcinomas, lymphomas, these clinically depressing names) has been described in carp, codfish, skate rays, pike, perch, and other fishes. Trout, like people, get liver cancer from a carcinogen, aflatoxin, produced by the fungus *Aspergillus flavus*. Rumors that sharks don't get cancer led to a mass slaughter by entrepreneurs hawking cancer-fighting shark cartilage pills. But sharks do get cancer. None of the classes of the animal kingdom are exempt. Among reptiles, there are cases of parathyroid adenoma in turtles and of sarcoma, melanoma, and lymphatic leukemia in snakes. Amphibians are also susceptible to neoplasms, but some offer a strange variation on the theme. When injected with carcinogens, newts rarely develop tumors. They are more likely to react by sprouting a new, misplaced limb. This ability to regenerate body parts has been all but lost by other animals over the course of evolution. Could this be another clue to the origins of cancer—damaged tissues trying frantically to regrow themselves, only to find that they no longer know how?

None of these creatures walk, swim, or slither to a clinic seeking care. But from the haphazard sightings of naturalists and zoologists, patterns have emerged. Mammals appear to get more cancer than reptiles or fish, which in turn get more cancer than amphibians. Domesticated animals seem to get more cancer than their cousins in the wild. And people get the most cancer of all.

One afternoon during my roadtrip, I stopped for a while at the Dinosaur Journey Museum. Given the current state of science museums—so much show biz—I expected the place to be infested with animatronic dinosaurs and hands-on exhibits resembling video games. But plenty of good science was there. I peeked through the picture windows of the Paleo Lab, where live men and women sat on display, leaning over worktables and chipping embedded fossils from surrounding stone. I walked among reconstructed skeletons towering toward the ceiling—*Allosaurus, Stegosaurus*. I saw a neck vertebra from an *Apatosaurus* so large that without the label I wouldn't have guessed the rocky mass had once been living tissue. It was all impressive, but over the years I had seen enough dinosaur skeletons to feel a little jaded. It wasn't until I stopped at a display with a full-size outline of a *Brachiosaur*'s heart standing as high as my chest that I really felt how enormous these beasts had been.

I thought again about Rothschild's survey of dinosaur tumors. There is a close relationship between size and life span. Though there are exceptions, larger species tend to live longer than smaller ones, and by some reckonings, the largest dinosaurs had very long life spans—so much time and space for mutations to collect. Wouldn't that have made them highly susceptible to neoplasms? At least in the mammalian world the issue is not clear-cut, an observation that goes by the name of Peto's paradox. It was named for Sir Richard Peto, an Oxford epidemiologist. He was puzzled that large long-lived creatures like elephants don't get more cancer than small short-lived creatures like mice. The mystery was succinctly posed in the title of a paper by a group of biologists and mathematicians in Arizona: "Why Don't All Whales Have Cancer?" Except for belugas in the polluted St. Lawrence estuary, whale cancer appears to be uncommon. For mice the cancer rate is high.

At first that didn't seem so strange. There is an inverse correlation between life span and pulse rate. During a typical lifetime an elephant and a mouse will each use up roughly a billion heartbeats. The mouse will just do it much faster. With a metabolism on so high a burn, it seems sensible that mice might get more cancer. But what is true for the mouse is not true for other tiny mammals. Birds, despite their frenzied metabolic rate (a hummingbird's heart can beat more than a thousand times a minute) appear to get very little cancer. If you graph mammalian size against cancer rate there is no telltale sloping line, just a scattering of dots. In our ignorance, each species seems like an exception.

Scientists have proposed several reasons for why cancer doesn't correlate smoothly with size. While larger animals may indeed get more mutations, they might also have evolved more effective means for repairing DNA, or for warding off tumors in other ways. The authors of the Arizona paper suggested how that might occur: hypertumors. Cancer is a phenomenon in which a cell begins dividing out of control and accumulating genetic damage. Its children, grandchildren, and great-grandchildren go on to spawn broods of their own—subpopulations of competing cells, each with a different combination of traits. The stronger contenders—those that have evolved an ability to grow faster than the others or to poison their neighbors or to use energy more efficiently—will gain an upper hand. But before they can dominate, the authors proposed, they might become susceptible to "hypertumors": clusters of weaker cancer cells opportunistically trying to latch on for a free ride. These parasites would sap energy continuously, destroying the tumor or at least keeping it in check. In large, long-lived animals cancer develops gradually enough for the leeches to form. They may indeed get more tumors, but they are much less likely to grow to a noticeable size. Cancer that can get cancer. For all the time I'd spent immersing myself in the literature, this was the first I had heard of that.

That still left me wondering about the hummingbirds, and a footnote in the paper about Peto's paradox led me to yet another of cancer's mysteries. It is well known to zoologists that virtually all mammals, no matter how tall or short, have precisely seven vertebrae in their necks: giraffes, camels, people, whales. (Manatees and sloths are exceptions.) Birds, amphibians, and reptiles are not bound by the rule—a swan can have twenty-two to twenty-five neck vertebrae. They also appear to get less cancer. Frietson Galis, a Dutch biologist, thought there must be some kind of connection. She considered what happens in rare instances when fetuses sprout an extra rib right where the seventh vertebra would normally be. As a result, children born with the defect have only six vertebrae in their necks. They are also more likely to die from brain tumors, leukemias, blastomas, and sarcomas. Galis suggests that it is why variation in the number of neck vertebrae is slowly being weeded out of the mammalian population.

I spent my last night on the road in Vernal, Utah, where a giant pink *Brontosaurus* (I mean *Apatosaurus*) with long flirtatious eye-lashes held up a sign welcoming visitors. It was about nine o'clock and the town was already shutting down. I found a restaurant with a Wild West theme barely open on Main Street. After a long day of driving I was looking forward to a glass of wine. I tried to keep up with the latest studies on how this vice, in moderation, might conceivably be good for the circulatory system, staving off heart attacks and strokes. The most wishful research even suggested that the anti-oxidizing effects of the elixir might help suppress tumors and extend life. But the longer you live the more likely you are to get cancer. Every meal presents a calculus of probabilities: Alcohol increases the risk for some cancers (mouth, esophageal) but may decrease the risk for kidney cancer.

In a file on my laptop I had been keeping a list of headlines from recent news:

> *"Natural Compounds in Pomegranates May Prevent Growth of Hormone-dependent Breast Cancer"*
> *"Green Tea Could Modify the Effect of Cigarette Smoking on Lung Cancer Risk"*
> *"Soft Drink Consumption May Increase Risk of Pancreatic Cancer"*
> *"Bitter Melon Extract Decreased Breast Cancer Cell Growth"*
> *"Seaweed Extract May Hold Promise for Non-Hodgkin's Lymphoma Treatment"*
> *"Coffee May Protect Against Head and Neck Cancers"*
> *"Strawberries May Slow Precancerous Growth in Esophagus"*

I knew by now that the effects, if real, would be minuscule. How can anyone sensibly weigh the trade-offs, based inevitably on imperfect information—on findings that could be overturned tomorrow?

The carcinogenic effects of red wine turned out not to be an issue that night. This was Utah and there was nothing alcoholic on the menu. My fried chicken cutlet sandwich was washed down with lemonade made with powder from a jar and tap water. Back at my room at the Dinosaur Inn (guarded over by another smiling *Apatosaurus*), I thought again about those layers extending miles and millennia below me. Someday more layers would pile on top of us, and I wondered how much cancer would be there. It had been seven years almost to the day since Nancy, the woman I was married to, was diagnosed with a rabid cancer that sprouted for no good reason in her uterus and burned like a flame along a wick down the round ligament and into her groin. She lived to tell the tale, but ever since, I have been wondering how a single cell minding its own business can transmogrify into a science fiction alien, a monster growing within.

## Chapter 2

# Nancy's Story

She always ate her vegetables. Obsessively, it sometimes seemed. Breakfast, lunch, dinner, throughout the day she would keep mental count. Never mind if it was 10:30 p.m., halfway through a *Simpsons* episode or a DVD. If she hadn't consumed two or three servings of vegetables (some green, some yellow) and three or four servings of fruits, nuts, grains—whatever the food pyramidologists were recommending—she would slice up an apple or open a bag of carrots.

In the spirit of Pascal's wager (there is no downside to believing in God), none of this probably hurt. It is often said that two-thirds of cancer cases are preventable—one-third by eliminating smoking and the other third by getting more exercise and eating healthier meals. But the evidence associating any particular diet with cancer is discouragingly thin. We were told, Nancy and I, to eat our spinach because it is rich in folates, and folates are a crucial ingredient used by cells to synthesize and repair the intertwining helices of DNA. That sounds great in theory, but the argument is weak at best that consuming more folate reduces the risk of three of the most common cancers: colorectal, breast, and prostate. For breast cancer, the effect, if there is one, may be primarily of benefit to alcoholics. Other research suggests that too much folic acid (the synthetic form of folate found in vitamin pills) can increase cancer risk. Once a neoplasm has taken root, extra doses might even accelerate its growth, adding fuel to the fire. Some cancers are combated by administering antifolates, which are among the oldest chemotherapeutic drugs. The most persuasive reason for eating spinach is that, sautéed with garlic or tossed in a salad, it tastes so good.

Just as dubious is the mythology surrounding antioxidants like vitamins C and E, which are consumed in fruits, vegetables, and pills and smeared on the face in the form of antiaging cosmetics. The hope is to counteract free radicals—products of cellular combustion that eat at the insides of cells. It is far from clear that the body needs help on that front. To blunt the impact of free radicals (the name conjures images of bomb-throwing anarchists), living cells come equipped with a built-in system of antioxidizing mechanisms, a finely strung molecular web crafted over the eons since life began. That is not the kind of thing you want to mess with. And no creature would want to eliminate free radicals. They are scavengers that prevent the inevitable accretion of cellular poisons, garbage collectors for the cells. Beta-carotene, an antioxidant that gives carrots, mangoes, and papayas their color, has been promoted as having anticarcinogenic powers. But in a clinical trial in Finland, smokers given beta-carotene supplements were more likely to get lung cancer. A similar trial in the United States was curtailed at an early stage when it also appeared that the supplements were increasing the risk of the disease. "To go beyond the bounds of moderation is to outrage humanity"—Pascal again—and to outrage our cells.

These days grocery store packaging has descended to a new level of detail, luring shoppers with produce and other goods rich in phytochemicals, naturally occurring ingredients in plants reputed to help detoxify carcinogens, repair DNA damage, or otherwise discourage cells from going wild. Lycopene, quercetin, resveratrol, silymarin, sulforaphane, indole-3-carbinol—they go in and out of style.

In a laboratory dish these substances might affect biochemical pathways believed to be involved in the numbingly complex processes of carcinogenesis. Far less clear is whether consuming more of them actually prevents anyone from getting cancer. Unless a person is severely malnourished there is little reason to believe that a shortage of any specific molecule is throwing the cellular processes seriously out of whack. You can hedge your bets by taking multivitamins, but the evidence here is also meager. If life were so delicate we probably wouldn't be here worrying about what we eat.

There is so much that science doesn't know about the molecular clockworks, and it is possible that substances in fruits and vegetables confer synergistic advantages whose logic is yet to be uncovered. Throughout the 1990s, the news was filled with reports of miraculous anticarcinogenic effects from consuming nature's bounty. The National Cancer Institute began pushing its 5 A Day program. Eat that many servings of fruits and vegetables and you would be a long way toward beating the odds against cancer.

The evidence, alas, came mostly from case control studies in which people with and without cancer were asked to remember what they ate. Epidemiological studies like these are prone to error. Grasping to explain their predicament, cancer patients might be more likely to overestimate how badly they neglected their diets, while healthy people might remember eating more fruits and vegetables than they really did. Since cancers can take decades to develop, great feats of memory are required. Skewing things further, those most likely to volunteer for the control group may be relatively affluent health-conscious citizens who, in addition to eating nutritious meals, exercise more often and are less likely to binge on alcohol or cigarettes. A good study will try to strike a balance between the cases and the controls, but the best that retrospective epidemiology can do is hint at associations to be investigated more rigorously. In prospective cohort studies, large groups of people—the cohorts—are followed for years and interviewed regularly to see if patterns emerge among those who do and do not get cancer. Though these too suffer from biases, the evidence is considered stronger than for retrospective epidemiology. The largest prospective study on diet and health has found so far that eating fruits and vegetables has, at most, a very weak effect on cancer prevention. There are suggestions of possible benefits with a few cancers, but nothing that has lived up to the earlier hopes.

We were told to eat our fiber, and when Nancy went shopping she would bring home breakfast cereals that tasted like pieces of the cardboard box. Intuitively it made sense. You could imagine all that fiber scouring clean your intestines on its way through the digestive tract. Fiber was also said to nurture a mix of bacteria that reduces the risk of colon cancer. The case for fiber may be a little stronger than for other foods, but the evidence has been controversial. One big prospective study found an association while another did not.

This all might be less ambiguous if foods could be subjected to the same kind of rigorous trials used to test new drugs. A large group of people is randomly assigned to an experimental group, which receives the treatment, or a control group, which does not. In the end the results are compared. But these studies are rare in cancer nutrition research. It is hard enough to force people to arbitrarily eat or not eat a certain food. And to clinch the case, the enforcement would have to last for the decades it can take a cancer to develop. When a four-year controlled trial was carried out with a low-fat diet that was high in fiber as well as fruits and vegetables it found no evidence of a reduction in colorectal polyps, which are precursors to colon cancer. Another randomized trial of about equal duration found that a fibrous diet had no effect on the recurrence of breast cancer.

Reading these less than ringing endorsements, I was reminded of the biochemist Bruce Ames, who has reported that brussels sprouts, cabbage, broccoli, cauliflower, and other staples of the farmer's market contain naturally occurring carcinogens, built-in pesticides like the ones that might have killed the poor *Edmontosaurus*. People apparently don't ingest these foods in quantities that would cause a public health problem—or maybe we have acquired a natural resistance. But how did the superstition arise that plants have the opposite effect, conferring us with the power to beat back cancer? Fruits and vegetables evolved to promote their own propagation. Then people started eating them.

There was nothing very rigid about Nancy's dietary pursuits. We both loved steaks and hamburgers, but we tried to moderate our consumption. Here the science sounds a little more persuasive. If the epidemiology can be believed, eating a lot of red meat every single day might have raised our chances of getting colorectal cancer during the next decade by as much as a third—from 1.28 percent to 1.71 percent. But given those odds, cooking a giant steak on the weekend seemed worth the trade-off. For penance we would have fish sometimes. Knowing that it is rich in omega-3 fatty acids may have made the salmon and halibut we grilled more satisfying. But any firm connection between fish, fish oils, and colon cancer prevention has remained elusive.

Fruits, vegetables, fibers, fish—if nothing else, loading up on these foods promised to reduce one's intake of mammalian fat. Yet even that has come under challenge as a serious cancer risk, and it is possible that sugar may pose a greater danger by increasing blood insulin levels and stimulating the growth of tumors. In the end, it probably doesn't matter so much what you eat as how much. Obesity—like old age, sunlight, radioisotopes, and cigarettes—has joined the short list of unambiguous instigators of cancer. Conversely there is evidence that caloric restriction reduces the likelihood of cancer. You lower your metabolism. Like a lizard.

Nancy included a variety of vegetables and fruits in our diet mostly because she liked them. But she had reason to worry more than some others about cancer. Her mother had suffered through a mastectomy and chemotherapy shortly before we married. After sixteen years of slumber, the cancer returned. We didn't know if her breast cancer was among those linked to a familial genetic defect. If so Nancy might have inherited a susceptibility, though not a fate.

She had other risk factors. She was forty-three and we had no children, a source of constant contention. The less frequently a woman is pregnant, the more monthly menstrual cycles she endures. With each period a jolt of estrogen causes cells in the uterus and mammary glands to begin multiplying, duplicating their DNA— preparing for the bearing and the nursing of a child that may not come. Each menstrual cycle is a roll of the dice, an opportunity for copying errors that might result in a neoplasm. Estrogen (along with asbestos, benzene, gamma rays, and mustard gas) is on the list of known human carcinogens published by the federal government's National Toxicology Program.

Women these days are also exposed to more monthly doses of estrogen because they are beginning to menstruate at much earlier ages, possibly increasing the risk of breast cancer. A few scientists blame the change on bisphenol A—a chemical in plastic bottles that mimics estrogen—but a more widely accepted explanation involves nutrition. With more food to eat, girls mature more rapidly, accu- mulating fat, and that may serve as a signal that the body is healthy enough to begin ovulation. Over a century the age of menarche, when menstruation begins, has dropped in the Western world from about seventeen to twelve. At the same time women are spending less of their fertile life either pregnant or nursing a child. Lactation also appears to hold estrogen in check. The result of all this is that a teenager today may have already experienced more menstrual cycles than her grandmother did during her entire life.

There are other risks in being female. Hormone therapies, admin- istered during menopause or pregnancy, have been associated with some cancers. And obesity, especially in older women, can increase estrogen along with cancer risk. But none of this is straightforward. Strangely enough, excess body fat can actually reduce the chances of premenopausal women getting breast cancer. And while oral con- traceptives may slightly raise the odds for cancer of the breast, they appear to reduce the risk of getting ovarian and endometrial cancer. Nancy wasn't using birth control pills and she was far from being overweight, but she worried, just a little, about another factor: the wine we liked to have with dinner. Alcohol might also tip the hormonal scales and has been associated for entirely different reasons with digestive cancers. Snuffed out by alcohol, epithelial cells lining the esophagus must be replaced—more DNA to be duplicated, more chances for error. There is evidence linking alcohol to liver cancer, but more certain is the risk from hepatitis viruses or long-term exposure to aflatoxin, a poison produced by funguses that can invade peanuts, soybeans, and other foods.

You could live your life with a calculator. Consuming two or three drinks a day might increase breast cancer risk by 20 percent. That is not as bad as it sounds. The chance that a woman between the ages of forty and forty-nine will get the cancer is 1 in 69, or 1.4 percent. Alcohol consumption would raise that to 1.7 percent. Even tallness is a risk factor. (Nancy was just five foot three.) An analysis of data from the Million Women Study found that every four inches over five feet increased cancer risk by 16 percent. A clue to the mechanism may be found in Ecuadoran villagers with a kind of dwarfism called Laron syndrome. Because of a mutation involving their growth hormone receptors, the tallest men are four and a half feet and the women are six inches shorter. Life is not easy for them. The children are prone to infections and adults frequently die from alcoholism and fatal accidents. But they hardly ever get cancer or diabetes, even though they are often obese.

When you're healthy and cancer remains an abstraction, enumerating life's hazards can be reassuring. Neither of us were smokers, in whom cancer risk is measured not in small percentages but in factors of ten to twenty. A twentyfold greater chance of getting lung cancer—nothing sounded subtle about that. From all the public service announcements and scary warning labels, I assumed that a large proportion of smokers must die that way. It was surprising to learn that the figure is more like 1 in 8. With a statistic like that, so many details are washed over. Surely the odds are far worse for a lifetime chain-smoker. In search of an answer I came across the online Memorial Sloan-Kettering cancer prediction tool. I plugged in some numbers. A sixty-year-old man who had smoked a pack a day since he was fifteen and now plans to give up cigarettes will have a 5 percent chance of getting lung cancer in the next ten years—and a 7 percent chance if he doesn't quit. I thought the odds would be so much worse. If the man is seventy and has smoked three packs a day, the risks are 14 percent and 18 percent. That still leaves heart attacks, strokes, chronic bronchitis, emphysema, and other cancers—a variety of ways to die. Smoking damages health and lowers longevity. But when you hear those stories about the uncle who smoked like a chimney every day of his life and never got lung cancer—that is the norm and not the exception.

Geography also plays into carcinogenesis, and there were dangers involved with living in Santa Fe, New Mexico, a place we loved for its stark juxtapositions. The semiarid plains giving sudden rise to 12,000-foot peaks. The old Spanish families sharing the same dirt street with artists and college professors. And there was the cool, dry, high-altitude air. It was too dry at times, and some summers we would anxiously watch smoke plumes billowing from distant forests. Ashes would fall from the sky, and the sun would set blood orange like images from Revelation. In the night the mountains glowed and erupted in plumes of fire. One of the fires swept through parts of Los Alamos. A study later concluded that the radiation spread by scorching the laboratory grounds posed one-tenth the risk of the naturally occurring radionuclides released by the burning pines. Good news, I guess—except for knowing that every forest fire may pose a measurable risk from nature's own fallout.

Santa Fe is nearly a mile and a half in altitude, so there is that much less atmosphere cushioning skin and eyes from solar rays. Sweeping from red toward blue on the spectrum, the frequency of light increases. The higher the frequency, the higher the energy, and by the time you get much beyond violet there is enough energy to break molecular bonds, to mutate DNA. Many times every summer a double rainbow would arch over Talaya, the conical peak at

Santa Fe's eastern edge. I was almost sure I could see, barely visible at the underside of the arc, a shimmering band of deadly ultraviolet. Beneath that would be colors our eyes don't know: x-rays and gamma rays. Sunlight is dangerous stuff. Yet there is some evidence, weak and conflicting, that the vitamin D it helps generate in the body lowers the odds for colorectal cancer—while raising the risk for cancer of the pancreas. At least among male Finnish smokers.

The assaults came from above and from below. As in so many parts of the country, the granitic soils our neighborhood was built on contained tiny amounts of naturally occurring uranium. Uranium-238 decays, shooting out alpha particles to become thorium-234 and eventually radium and then radon, a radioactive gas that cannot be seen or smelled. Radon is considered a risk factor for lung cancer, occupying a distant second place behind cigarette smoking, and is being investigated for a lesser role in other cancers. It accumulates at a geological pace (the half-life of U-238 is more than 4 billion years, meaning that it would take that long for half a portion to decay). The gas itself lingers only a few days, breaking down into radioactive daughter particles and ultimately into minuscule traces of lead. But it is constantly being generated, and when I bought our house the inspector measured 5.4 picocuries per liter of air, a little above the Environmental Protection Agency's "action level" (4 picocuries per liter) at which a follow-up test was recommended and people were advised to consider radon mitigation with sealers, blowers, and vents. I began caulking floor cracks—Pascal's wager—which had the more tangible result of reducing the population of spiders and centipedes. I was soon diverted by other things. For someone who never smoked, 4 picocuries per liter poses a lifetime risk of dying from lung cancer of about 7 in 1,000—less than 1 percent—and that assumes constant exposure, as if you spent your life indoors like a shut-in or a kidnap victim.

We lived near no industrial sites and Los Alamos, the Atomic City, was twenty-five miles away, on the far side of the Rio Grande Valley. In the early 1990s, an artist living there had reported what appeared at first to be a high number of brain tumors in his neighborhood. State health officials investigated. During the previous five years there had been ten cases in the county instead of the six that would be expected from state and national averages. But the numbers were too small to be meaningful, and epidemiologists concluded that there was no way to distinguish the increase from what could have arisen through chance. There was nothing unusual or alarming, they said. If you stepped back and examined the world at large you would find similar bunchings in space and time, but you would have no reason to assume that they pointed to an underlying cause. Epidemiologists talk about the Texas sharpshooter effect. Blast a barn door with a shotgun and then find the holes that are closest together. Draw a target around them and it looks like you hit a bull's-eye. As soon as it peaked, the brain cancer rate fell and then zigzagged around normal. The Los Alamos investigators had also found a blip in thyroid cancer. But again the numbers were small—a total of 37 cases over twenty years in a population of 18,000—and in the following years they too declined. A public health assessment concluded that residents were receiving no harmful exposures from chemical or radioactive contamination whether from the water, soil, plant life, or air.

In thinking about exposures, there was also the past to consider. Nancy had grown up in New York, on Long Island, where in the early 1990s the suburbs began to reverberate with fears of a breast cancer epidemic. When a friend or family member is struck out of the blue with a malignancy, the mind becomes magnetized, pulling in specks of data. There is that woman down the street who was also diagnosed with breast cancer. And the sister-in-law in the next town and the wife of the man at the office. The brain, built to seek patterns, insists on connections. The Long Island cancer cluster was born.

And so you start looking for a reason, a source, the spider crouched at the center of the web. Was it the Brookhaven National Laboratory, with its particle accelerators and research reactors? Or the pesticides and weed killers used in the old days when the island was mostly farmland—and, more recently, to maintain all of those flawless Long Island lawns? Or the DDT that had been sprayed to control mosquitos? Was it the high density of power lines in an area hungry for electricity?

The worry and fear—so reasonable, so understandable, so very human—lapsed sometimes into hysteria, which lapsed into paranoia. One activist ominously alluded to "a type of population control," as if Long Islanders were being exterminated, willfully or by neglect, by the wholesale damaging of their genes. The politicians had to listen and Congress mandated a study. A decade later the National Cancer Institute issued its $30 million report. The incidence of breast cancer in Nassau and Suffolk Counties was slightly higher than for the United States as a whole. But the same was true for much of the urban Northeast—a clue that anything the cancers might have in common was very diffuse. The cluster was more like a sprawl.

No link was found between pollutants and breast cancer. If there were more cancer cases on Long Island, the study concluded, the causes were probably socioeconomic. There may also have been a genetic factor. Many Long Island women were from Ashkenazi Jewish families, which show a propensity for breast cancer. But the likeliest culprit was the relatively affluent suburban lifestyle. Long Islanders were apt to eat richer diets, to be overweight, to bear fewer children, and to live longer—the median age for diagnosis of breast cancer is sixty-one. Long Islanders were better educated than average and therefore more likely to get frequent mammograms, discovering tiny, slow-growing, possibly harmless neoplasms that are treated, just to be sure, and recorded in the statistics. A woman living in a shack in Appalachia might carry these "in situ" carcinomas to the grave, dying beforehand of something else.

These are not the kinds of reasons people want to hear—that their cancer might have been prevented had they chosen to forgo careers and, like the squirrels and foxes, be pregnant all of the time. That they might have enjoyed too many good meals and gained too much weight. That their lump mastectomy might have been unnecessary. "Blaming the victim," some activists complained, and one of them dismissed the report altogether: "We certainly believe there is an environmental connection, and we don't have to have proof to say what it is."

Everyone has risk factors for almost every cancer, and they take on significance only in retrospect. One day our neighbor Vivian, happily working at home as a translator of scientific documents, "presented," as they say on grand rounds, with cancer of the ovaries. She died on an Easter Sunday, and the next thing we knew we were sitting at her memorial service. She was married to a mathematician. There was no mention of God. Around the same time, Susan, a former girlfriend of mine and a colleague from the journalism world, also died of ovarian cancer. Both she and Vivian were childless. But there was also Mrs. Trujillo across the street, a mother well beyond middle age who died of the same thing. All of us acquire our own personal cancer clusters, and a mental file of anecdotal evidence as unreliable as it is impossible not to deep down believe.

When Nancy's cancer came we didn't know if it had started in her ovaries, her breasts, her uterus, her lungs. For the longest time (weeks—the clock was ticking so slowly) we didn't know where it was growing, only that it was shedding cancerous cells into her body. She had been visiting a girlfriend in San Diego and was doing sit-ups in a local gym when she noticed a lump on the inside of her right groin. The words "swollen lymph node"—like what you might get from a sore throat—leapt to mind.

Cat scratch fever, we quickly decided, after seeking reassurance from the Web. Weeks earlier, startled by a sudden sound, one of our cats had clawed her leg, and an immune response from an infection could have led to lymphatic swelling. That is what lymph nodes are for, to capture and neutralize immunological invaders. The human mind, ever hopeful, has a talent for absorbing aberrations.

The bump didn't go away. Her doctor thought it might be a hernia and recommended a consultation with a surgeon. But that didn't happen right away. A phone call from out east brought news that Nancy's father had suffered a hemorrhagic stroke—what a horrible year this was—and lay in intensive care at Stony Brook University Medical Center. The appointment with the surgeon was postponed and a flight to LaGuardia Airport was booked. Nancy called home that evening and told me about sitting at his bedside: his eyes, his smile, the grip of his hand, his obvious comprehension. He filled every cubic inch of her soul, except for a tiny space. The space got larger. In the days since her arrival, the lump had obstinately endured.

She didn't have to leave the Stony Brook campus for a medical consultation. The next time she called she was walking back to her car from an appointment at a clinic, past familiar buildings (she had taken a degree in biology there). Her voice was wavering just enough that I knew she was probably crying, or trying not to. The doctor had palpated the lump. It was not soft and round as it would be from an infection. It was not cat scratch fever. It had the hard, irregularly shaped feel of a malignancy. The look on his face told her that she almost surely had cancer. He recommended a needle biopsy—the sucking out of cells to see if they are malignant. She decided to come home for the procedure.

There are those times we all come to know when you are sitting in a hospital waiting room surrounded by other people—the older ones flipping through magazines, the younger ones staring into the bottom of their cell phones. I had been through that with my mother after her torn rotator cuff and when the second of her knees was replaced. I had been through it with Nancy for a detached retina after a horseback ride. I knew what to expect. Just when you think you cannot endure another minute, the surgeon walks in, her mask hanging around her neck. She is smiling, pleased to be giving you good news. This time that didn't happen. "We may be looking at a carcinoma," she said.

She had sent a sample of the lump downstairs to pathology for a quick look under a microscope. The misshapen cells resembled epithelial cells that form the lining of organs. But they had mutated enough to become less differentiated. They were losing their genetic identity. Reverting to this primitive state, cells bear a resemblance to those in an embryo—rapidly dividing, chameleon-like, and capable of doing almost anything.

The diagnosis would have to be confirmed in the laboratory. But there was little doubt about what was happening. I walked with the surgeon to the recovery room where Nancy lay in an anesthetized blur. I remember her smiling as the surgeon spoke, and I only realized later that she was barely absorbing the information. For the rest of the week I tried to be optimistic, and maybe I unintentionally misled her. My understanding was that the diagnosis was, say, 90 percent certain, that the lab report was a technicality, a way to be absolutely sure. I thought that was Nancy's understanding too.

A few days later I was upstairs in my office when the doctor called her to break the news. "Extensive metastatic adenocarcinoma, moderately differentiated." Adenocarcinomas are carcinomas of epithelial tissues that contain microscopic glands. They can arise in the colon, lung, prostate, pancreas, almost anywhere. I don't remember how I knew to walk downstairs. Or did she walk upstairs to me? I had never seen her so upset. She told me that she had hung up the phone and screamed. Somehow cancerous cells had gotten into her lymphatic system and lodged inside that node in her groin. But where in her body had the cells come from? It would be weeks before we knew. "Metastatic cancer with an unknown primary"—it seemed like the worst possible diagnosis. A tumor was single-mindedly growing, shedding more seeds, metastasizing. But no one knew where.

There were hints from the pathology report describing the character of the cells:

ESTROGEN RECEPTORS *Approximately 90% positive (favorable)*
PROGESTERONE RECEPTORS *Negative (unfavorable)*

The first line provided a scrap to hang on to. Since the growth of some cancers is driven by estrogen, it might be controlled by blunting its effect. The abundance of these receptors also helped narrow the diagnosis:

*Comment: The estrogen receptor positivity is consistent with an endometrial or ovarian primary rather than a gastrointestinal primary.*

So it was probably gynecological. The endometrium—the lining of the uterus—is a tissue of epithelial cells, which are vulnerable to carcinomas. I think Nancy had suspected something like that. About a year earlier she had been told by her doctor that she was experiencing unusually early menopause. The sign was irregular menstrual bleeding, and I still wonder why that was not taken as a warning, an occasion for more tests—whether the cancer might have been discovered then and treated before it had been allowed to spread.

*Comment: The tumor has micropapillary architecture suggestive of endometrial, ovarian or*

The rest of the sentence was cut off. Monkeys with typewriters recording your fate. And being sure to include the billing code.

The surgeon was so supportive, so sympathetic, so sisterly. On a follow-up visit she gave Nancy a hug. I think we both were stunned when her next move was to hand us a pile of paper—yellow, pink, blue—orders for various procedures. We were to take them to local clinics, stand in line, and apply for an appointment. The chain store imaging center across the street would do a mammogram, chest x-ray, and a CT scan of the abdomen and pelvis. The colonoscopy factory was booked solid, the surgeon said. Rather than insisting that a patient with metastatic cancer be accommodated—most people's colonoscopies are routine and could easily be rescheduled—she issued an order for a barium enema, an ancient, quicker, less definitive test. We asked about a referral to an oncologist. That would be premature, the surgeon told us, until we knew what kind of cancer this was. She actually said that.

What is a crisis for the patient is routine for the doctor, but this still seems to me like pure idiocy. We went from lab to lab, returning to pick up the results. The mammogram and chest x-ray were negative. The abdominal scan showed the liver, kidneys, pancreas, bowel, and lower lungs to be normal. So were the adrenal glands. A 1.3 centimeter nodule in the area of the spleen looked "to be a splenule only"—a benign mass that can sometimes be confused with a tumor. In the pelvic scan a cyst on the left ovary appeared "unlikely to be neoplastic" but the uterus and endometrium were "prominent" and there were benign fibroids. There was a "question of [a] small constrictive lesion of sigmoid colon." It was scary reading language whose nuances we were not attuned to understand. Especially disturbing were the results of a blood test: CA-125, a protein found in higher concentration with some cancers, was elevated. The test was far from conclusive—many other things can cause a high reading—but it hinted at the possibility of ovarian cancer, the kind that had killed our friend Vivian.

As we accumulated information we also made phone calls. I talked to a doctor at the Mayo Clinic in Scottsdale where I had splurged for my fiftieth birthday on an executive physical. She suggested the obvious: MD Anderson Cancer Center in Houston or Sloan-Kettering in New York. I contacted those places and—more importantly it turned out—found out who had treated Vivian. Her husband spoke highly of her oncologist, and when I called the doctor's office in Santa Fe, his secretary squeezed us in for an appointment.

Imagine a tall, thin, aging cross between Jimmy Stewart and John Wayne—I think he was wearing cowboy boots—ambling in and taking charge. He was reassuring in his casualness, someone who had seen it all. He leafed through the test results. "There's nothing here really." He said it was unlikely that an ovarian cancer would metastasize to an inguinal lymph node. He was puzzled by the order for a barium enema, which was scheduled for a few days later. "It's a useless test," he said. We explained about the long wait for a colonoscopy. He picked up the phone, called one of the physicians who owned the clinic, and we had an appointment two days later. "We are going to cure you," he said. That at least is my memory. Oncologists are not supposed to say that. It was encouraging that he didn't give a damn.

The results of the colonoscopy were negative, and the final step was a PET scan. Santa Fe had just gotten its own machine—it was no longer necessary to drive an hour south to Albuquerque—and Nancy was almost first in line. PET stands for positron emission tomography, a triumph of medical technology from the recondite world of particle physics. The patient fasts the night before so the body's cells are starving. When radio-tagged glucose is injected it is eagerly consumed. Malignant, rapidly dividing cells are especially voracious, concentrating the radioactive molecules. As these decay they shoot out positrons, particles of antimatter that collide with electrons and produce bursts of gamma rays. They strike a scintillator, which responds by emitting flashes of light. Nancy's lower uterus was glowing from the feasting of hyperactive endometrial cells, descendants of a single cell gone mad, a cell that had forgotten it was part of a community, that began running its own show—an isolated act of betrayal that has been played out again and again since the first archaic cells grudgingly agreed to surrender their autonomy for the advantages of living in a collective.

In the days after the diagnosis, I began reading about how this might have happened. To carry on in harmony, our cells constantly exchange chemical signals, conferring on when to start multiplying and creating new tissue. As each cell receives this information it responds by sending instructions to its nucleus, the central controller, for activating the appropriate combination of genes—pressing the right buttons, hitting an arpeggio of piano keys. A cancer cell is one that has cut itself out of the discussion, solipsistically deciding on its own. Random events—triggered by a cosmic ray, a carcinogenic chemical, or just plain dumb luck—must have altered the DNA inside one of Nancy's cells, causing it to lose touch. The trouble might have begun with a mutation to a gene that sends signals telling the cell that it is time to divide. Another mutation might have modified the molecular receptors that respond to the signals, causing them to become hypersensitive. Set on a hair trigger they fire prematurely. Either way, the cell begins multiplying more rapidly than its neighbors.

In fact these kinds of errors happen all the time. We usually don't get cancer because other genes react to sudden bursts of activity by reining in growth. But another mutation can cause that safeguard to fail. The nucleus of a cell is constantly receiving messages, weighing the evidence and deciding what to do next. The calculations depend on a tangle of molecular cascades—more things that can go wrong. And they do. All the time. The mistakes are caught and corrected. The DNA is repaired. If that fails, a cell can sense the inner turmoil and send itself suicide signals, killing itself for the common good. But another mutation can undermine that defense.

This is all usually described as though a single cell was sitting motionless and accumulating these defects over the years. I tried to imagine the process as it really is, unfolding dynamically. One hit causes a cell to start dividing repeatedly. Then one of its many progeny acquires another mutation, and its progeny acquire still more. The longer a lineage of cells has lived the more likely it is to have mutated to the brink. That still leaves another barrier against runaway growth: a counter that monitors and limits how many times a cell can divide. With the right mutation a cell can learn to continually reset the count and become immortal. Copying itself again and again, it produces a mass of mutant offspring, a tumor.

And that is still not enough to give you cancer. It takes more mutations for the cell to learn how to invade surrounding tissues, to become malignant rather than benign. Even so, the tumor can grow only so large—the size of the tip of a ballpoint pen—before it's starved for food or drowns in its own waste. For the tumor to continue expanding it must find a way to reach into the circulatory system and suck like a vampire.

With this infusion of nutrients the cells multiply more aggressively than ever, increasing the probability of more mutations—or adaptations, from the point of view of the evolving cancer cell. The phenomenon is what computer scientists call "random generate and test." With all the restraints removed, the genome spins out one variation after another—hopeful monsters trying to gain an upper hand. Some might learn to consume energy more efficiently, others to tolerate harsher environments or to suppress the immune system. Finally the fittest will set sail in the bloodstream or the lymphatic ducts and explore new ground.

As I thought about this I was pulled in opposite directions. With so many checks and balances, a person must be extraordinarily unlucky to get cancer. Then again, with so many things that can go wrong, it is amazing that cancer doesn't happen all the time.

## Chapter 3

# The Consolations of Anthropology

When Louis Leakey sat down to recount the discovery of what may be the earliest sign of cancer in the genus *Homo,* the first thing he remembered was the mud. It was March 29, 1932, midway through the third East African archaeological expedition, and it had rained so long and so hard that it took an hour to drive the four miles from the campsite in Kanjera, near the shore of Lake Victoria, to the Kanam West fossil beds. By the time he and his crew had slogged their way through they were covered with mud, and before long Leakey, who was just beginning an illustrious career as an anthropologist, was on hands and knees scouring the ground for newly exposed bones.

He was coaxing the remains of an extinct pig from the muck when one of his Kenyan workers, Juma Gitau, walked over with a broken tooth he had just extracted from a cliffside. *Deinotherium,* Leakey noted, a prehistoric elephant-like creature that roamed Africa long ago. Gitau went back to look for more, and as he was scratching away at the cliff face, a heavy mass of calcified clay broke loose. He chopped it with his pick to see what was inside: more teeth, but not *Deinotherium.* These looked like what a dentist might recognize as human premolars, still set in bone, yet they came from a layer of sediment deposited, Leakey believed, in Early Pleistocene time, about a million years ago.

Back at Leakey's home base at Cambridge University the Kanam mandible quickly became a sensation—"not only the oldest known human fragment from Africa," he proclaimed, "but the most ancient fragment of true *Homo* yet discovered anywhere in the world." It was radical enough in those days to claim that man had originated in Africa, rather than Asia, where primordial ancestors like Java man and Peking man had been discovered. They may have been of approximately the same age as Kanam man, but Leakey found their features to be more apelike in appearance. The Kanam mandible, to his eyes, showed more modern characteristics including remnants of a human-like chin—evidence that *Homo sapiens,* not just its slack-jawed cousins, was far older than previously believed. Differences in the shape of the teeth led Leakey to consider Kanam man a slightly different species: *Homo kanamensis.* It was, he insisted, the direct precursor of us all.

Like many of Leakey's enthusiasms this one proved controversial. One of his detractors thought the specimen looked *too* modern, that it was a more recent jawbone that had washed into much older surroundings. In later years anthropologists speculated that what Leakey called *Homo kanamensis* might actually be a more distant relative like *Australopithecus,* Neanderthal man, or *Homo habilis.* More recently others have come to believe that the specimen may be Middle to Late Pleistocene, which would make it no more than about 700,000 years old. Whatever its pedigree or precise age, Kanam man is no longer considered remarkable for its antiquity but for an abnormal growth on the left side of the jaw.

At the time of the discovery, it had seemed like a bother, detracting from Leakey's find. He was working in his rooms at St. John's College, Cambridge, carefully cleaning the specimen, when he felt a lump. He thought it was a rock. But as he kept picking he could see that the lump was part of the fossilized jaw. He sent it to a specialist on mandibular abnormalities at the Royal College of Surgeons in London, who diagnosed it as sarcoma of the bone.

There was also a thin fracture in the jaw, one that had occurred long enough before death to heal. That, the doctor surmised, may have been how the cancer had begun. Sensing the trauma, as bone cells somehow do, they had begun rapidly dividing, replacing dead tissue. And somewhere along the way—the odds are vanishingly small—this carefully controlled process had gone askew. More than enough new cells had been produced to heal the wound, but they didn't know when to stop. Because of some biological miscalculation, cells kept dividing and dividing, overflowing the crack. Plausible as it sounded, this was just speculation. Bone fractures have not been established as a trigger for osteosarcoma. Usually there is no obvious cause. However the cancer begins it often spreads to the lungs. If the diagnosis is correct—some have had their doubts—that may be what killed Kanam man.

I first came across a mention of the Kanam jaw in a history of cancer timeline somewhere on the Web. That sent me digging into Leakey's old books and papers, and after several e-mail exchanges I tracked down the fossil at the Natural History Museum in South Kensington, London, where it had been in storage for decades. As far as I could tell it had never been on display. The specimen had been removed from the shelf now and then to be examined. The anthropologist Ashley Montagu studied it in 1956, reporting that the tumor was so large and disfiguring that it was impossible to tell what Kanam man's chin had been like. Other anatomical details persuaded him, however, that the fossil was clearly human-like. Another anthropologist disagreed, concluding that what Leakey thought was a chin was part of the tumor.

And so the disputes began. A London oncologist, George Stathopoulos, ventured that the tumor might not be osteosarcoma but an entirely different cancer, Burkitt's lymphoma, a malignancy of the lymphatic system endemic among children today in central Africa, one that often damages bone. Others were not so certain. Osteomyelitis, a chronic infection, can also generate bony growths. But in his book *Diseases in Antiquity,* a standard reference on ancient pathology, Don Brothwell concluded that Kanam man's abnormality was too thick and extensive to be from an infection. Like Leakey's colleagues, he leaned toward a diagnosis of bone cancer. As recently as 2007, scientists scanning the mandible with an electron microscope concluded that the crack had indeed resulted in "bone run amok" while remaining neutral on the nature of the disease.

I wanted to see the specimen for myself, and on a spring day I arrived, as previously arranged, at the museum's staff and researcher entrance on Exhibition Road. The man at the guard desk called ahead to Robert Kruszynski, curator of vertebrate paleontology. "He asks that you meet him by the giant sloth." It was easy enough to find. Hunched on its hind legs, the creature's plaster cast skeleton towered over the heads of museumgoers as it prepared to chomp at the top of an artificial tree. It had been standing that way for 161 years, when it was assembled from the bones of two or more South American specimens and put on display. Behind me was a wall of *Ichthyosaurus* fossils, mounted in glass cases. As I examined them, marveling at how the same bony architecture runs throughout the vertebrate world, a door opened in the corner of the hall. Mr. Kruszynski came out to greet me and then led me into the museum's inner sanctum.

Waiting for me on a table by a window was the brown cardboard box he had retrieved from the museum stores. The handwritten label identified the contents:

M 16509

KANAM MANDIBLE.

"M" stood for "mammal." In the upper right-hand corner of the label were two colored stickers—a red sunlike symbol and below that a blue star—indicating that the specimen in the box had been analyzed at various times by radioassay and x-rays. Mr. Kruszynski carefully removed the lid. Inside was a smaller box, fashioned from balsa wood and cardboard and covered with a glass lid, and inside that was the Kanam jaw.

He placed it on a padded mat, two layers thick to cushion it from the hard surface of the table. "All yours to look at," he said, and went off to search for another fossil I hoped to see: a femur retrieved from an early medieval Saxon grave in Standlake, England, with an enormous growth that had also been diagnosed as a cancerous bone tumor.

I had thought I would be content just glimpsing the Kanam jaw. I never expected to be left alone with it and to be able to hold it in my hand. It was dark brown and unexpectedly heavy and dense. That shouldn't have been surprising. It was a rock really, petrified bone. Once it had been part of a prehistoric man, or a protoman. Two yellowed teeth were still in place, and there was a deep hole where the root of another tooth had been.

Just below that, on the left inside curve of the jaw, was the tumor. It was bigger than I had expected, reminding me perversely of a type of candy from my childhood called a jawbreaker. There was also a slight swelling on the outside of the jaw, and I could understand how people might argue endlessly over whether it was part of a tumor or a chin. I could see where Leakey had sliced through the mass (some of his colleagues considered this sacrilege) to remove a section for further analysis. I could almost picture the rest of the head, its vacant eyes pleading for relief from inexplicable pain.

Mr. Kruszynski returned half an hour later to see how I was doing with the fossil. "Don't bring it too close to the edge," he warned. I suddenly realized that the protective pad on the table was sloping toward my lap and how easily a sudden movement might have sent the Kanam mandible dropping onto the linoleum floor.

In the end Mr. Kruszynski was unable to find the cancerous femur I'd inquired about. "For another time," he said. The museum stores were undergoing a renovation, he explained, and the bone had apparently been mislaid along with the rest of the skeleton—all except for the skull. He pulled it from its box and let me hold it for a minute—so lightweight compared to petrified bone—then escorted me back across the barrier to the public portion of the museum.

Hundreds of visitors of all ages coursed through the hallways. Some of them inevitably would get cancer, or they would love somebody who did. I wondered if anyone had been there for Kanam man.

Not much has been written about the obscure discipline of paleo-oncology. Although research had gone on sporadically for decades, the word was introduced to literature only in 1983 when a small group of Greek and Egyptian oncologists (from the Greek *onkos,* meaning "mass" or "burden") began planning a symposium on human cancer in earlier times. The gathering took place the following year on a voyage between the island of Rhodes and the island of Kos, where Hippocrates was born. What emerged was an elegantly published, sparsely printed little book, *Palaeo-Oncology.* I felt lucky to find a copy on the Internet for one hundred dollars. Its fifty-eight pages are bound in a blue cover with gilded print, and below the title is a drawing of a crab. "Crab" in Greek is *karkinos,* and Hippocrates, in the fifth century B.C., used the word—it became the root of "carcinogen" and "carcinoma"—for the affliction whose Latin name is cancer.

It is not clear exactly why he chose the name. Some six hundred years later, Galen of Pergamon speculated on the etymology: "As a crab is furnished with claws on both sides of its body, so, in this disease, the veins which extend from the tumour represent with it a figure much like that of a crab." The story is repeated in almost every history of cancer. Very few tumors, however, look like crabs. Paul of Aegina, a seventh-century Byzantine Greek, suggested that the metaphor was meant to be taken more abstractly: "Some say that [cancer] is so called because it adheres with such obstinacy to the part it seizes that, like the crab, it cannot be separated from it without great difficulty." The word *karkinoi* was also applied to grasping tools like calipers.

All but forgotten is a very different derivation from Louis West-enra Sambon, a British expert on parasitology who, before his death in 1931, turned his attention to the study of cancer. There is a parasite, *Sacculina carcini*, that feasts on crabs in a manner eerily similar to the feasting of a cancerous tumor. The process was described in 1936 in a report by the pathologist Sir Alexander Haddow to the Royal Society of Medicine:

> [I]t attaches itself to the body of a young crab, and casts off every part of its economy save a small bundle of all-important cells. These penetrate the body of the host and come to rest on the underside of the latter's intestine, just beneath the stomach. Here, surrounded by a new cuticle, they shape themselves into the "sacculina interna," and like a germinating bean-seedling, proceed to throw out delicately branching suckers which, root-like, extend through every portion of the crab's anatomy to absorb nourishment. Growing in size, the parasite presses upon the underlying walls of the host's abdomen, causing them to atrophy, so that when the crab moults, a hole is left in this region corresponding in size to the body of the parasite. Through this opening the tumour-like body finally protrudes and becomes the mature "sacculina externa," free to deliver the active young into the open waters.

Long before the days of Galen, disciples of Hippocrates, dining on crabs, may have noticed the similarities between the way the parasite overtakes its host and the way a cancer metastasizes.

Whatever the reason for the name, ancient Greek texts describe what sound like cancer of the uterus and the breast. Driven by a belief in sympathetic magic, some physicians would treat a tumor by placing a live crab on top of it. They also recommended powders and ointments (sometimes made from pulverized crabs) or cauterization (burning closed the ulceration). As for patients with internal tumors, Hippocrates warned that they might best be left alone: "With treatment they soon die, whereas without treatment they survive for a long time." The principle is part of the Hippocratic oath: First do no harm.

With Galen the references become even sharper. He wrote an

entire book about tumors and included malignancies in a category of growth called *"praeter naturam"*—preternatural, meaning outside of nature. Carcinoma, he wrote, is "a tumor malignant and indurated, ulcerated or non-ulcerated." He found breast cancer to be the most common and especially prevalent after menopause. (In contradiction to what modern oncologists believe, he wrote that women who regularly menstruate don't get cancer.) He writes about uterine, intestinal, and anal cancer, and cancer of the palate. Sometimes he, like other Greek writers, uses the word *therioma,* "wild beast," to mean malignant. "The early cancer we have cured, but the one that rose to considerable size, without surgery, nobody has cured."

The medieval surgeon Abu al-Qasim al-Zahrawi was no luckier: "When a cancer has lasted long and is large, you should not come near it. I have never been able to save any case of this kind, nor have I seen anyone else who has been successful."

It is not so different now.


There is something comforting about knowing that cancer has always been with us, that it is not all our fault, that you can take every precaution and still something in the genetic coils can become unsprung. Usually it takes decades for the micro damage to accumulate—77 percent of cancer is diagnosed in people fifty-five or older. With life spans in past centuries hovering around thirty or forty years, finding cancer in the fossil record is like sighting a rare bird. People would have died first of something else. Yet in spite of the odds, cases continue to be discovered, some documented so vividly that you can almost imagine the ruined lives.

After my visit to London I received from the Natural History Museum photographs of the Saxon skeleton whose tumorous femur I had hoped to examine. I had read that the growth was large—10 inches vertically by 11 inches horizontally—but I was astonished to see what looked like a basketball grafted onto the young man's leg. The tumor shows a sunburst pattern that pathologists recognize as a sign of osteosarcoma. They see it most often in adolescents whose limbs are undergoing hormone-induced spurts of growth—more evidence for one of cancer's few established rules: The more frequently cells are dividing, the more likely mutations will occur. The right combination will lead to a malignancy. Osteosarcoma is so rare that one would have to comb through the bones of tens of thousands of people to find a single example. Yet ancient cases continue to turn up.

There were signs of the cancer in an Iron Age man in Switzerland and a fifth-century Visigoth from Spain. An osteosarcoma from a medieval cemetery in the Black Forest Mountains of southern Germany destroyed the top of a young child's leg and ate into the hip joint. Bony growths inside the roof of the eye sockets indicated anemia, which may have been an effect of the cancer. The authors of the report speculated on the cause: contamination from a nearby lead and silver mine. Cancer is especially hard to accept in children, even in one from nine centuries ago, and the paper ended with a poignant note: "The tumour would certainly lead the child to die a painful death." Though child mortality was very high in those days, the authors noted, children who made it past the first few years might live into their forties. But not this time. "The flame of life in the affected child was extinguished just when the child had survived the first years of infant excess mortality."

Maybe it helped to believe there was a reason—metallic poisoning from a mine. But no one knows what causes osteosarcoma. Then, as now, a few cases probably were hereditary, traced to chromosomal abnormalities. In modern times speculation turned for a while to fluoride-treated water and, more plausibly, radiation—therapeutic treatments for other disease or exposure to radioactive isotopes like strontium-90, which is spread by nuclear fallout. Strontium sits just below calcium in the periodic table of elements and imitates its behavior, incorporating itself tightly into bone. But most often osteosarcoma strikes for no apparent reason, leaving parents grasping to understand what remains as inexplicable as a meteor strike.

Another malignancy, nasopharyngeal carcinoma, which affects the mucous membrane in the nose, can scar adjacent bone, and signs of it have been found in skeletons from ancient Egypt. One woman's face had been all but obliterated, and I tried to imagine her stumbling through life. "The large size of the tumor, which caused such extensive destruction, suggests a relatively long-lasting process," observed Eugen Strouhal, the Czech anthropologist who documented the case. "The patient seems to have survived for a considerable time, and doubtless had pain and other symptoms. Survival would be impossible without the help and care of the patient's fellow-men." Here was another case where the horrors of cancer punched through the flat veneer of scientific prose.

Multiple myeloma, a cancer of plasma cells in bone marrow, can leave skeletal marks. Traces were found in the skull of a woman who lived in medieval times. Plasma cells are part of the immune system and when behaving normally they produce antibodies called immunoglobulins. In multiple myeloma, one type is generated at the expense of the others. A chemical test found antibodies that the researchers considered confirmation of the disease.

Osteosarcoma, nasopharyngeal carcinoma, multiple myeloma—these are primary cancers, those found at the site of origin. They are debilitating enough. Most skeletal cancers by far come from metastases originating elsewhere. They also show up with greater frequency in the fossil record—and with devastating results. Metastatic bone cancer has been discovered in Egyptian tombs, in a Portuguese necropolis, in a prehistoric grave in the Tennessee River valley, in a leper skeleton from a medieval cemetery in England. Buried near the Tower of London the skeleton of a thirty-one-year-old woman was marked with metastatic lesions. We even know her name from a lead coffin plate: Ann Sumpter. She died on May 25, 1794.

In 2001 archaeologists excavated a 2,700-year-old burial mound in the Russian republic of Tuva, where nomadic horsemen called the Scythians once thundered across the Eurasian steppes, their leaders exquisitely dressed in gold. Digging down through two wooden ceilings, the scientists came upon a subterranean chamber. Its floor, covered with a black felt blanket, cushioned two skeletons. Crouched together like lovers, both man and woman wore what remained of their royal vestments. Around the man's neck was a heavy band of twisted gold decorated with a frieze of panthers, ibex, camels, and other beasts. Near his head lay pieces of a headdress: four gold horses and a deer. Golden panthers, more than 2,500 of them, bedecked his cape. His riches couldn't save him. When he died—he appeared to have been in his forties—his skeleton was infested with tumors. A pathological analysis, including a close look with a scanning electron microscope, concluded that the nature of the lesions and the pattern of their spread were characteristic of metastatic prostate cancer. Biochemical tests revealed high levels of prostate-specific antigen, or PSA. For all the false positives these tests can produce, this result was apparently genuine.

Metastasizing prostate cancer has been diagnosed in the partially cremated pelvis of a first-century Roman and in a skeleton from a fourteenth-century graveyard in Canterbury. While prostate cancer tends to be osteoblastic, adding unwanted mass to the skeleton, breast cancer is osteolytic, gnawing mothlike at the bone. Of all cancers, prostate and breast show the strongest appetite for skeletal tissue. Depending on the gender of the victim they are the first choice for diagnosis when bone metastases are found.

A middle-aged woman with osteolytic lesions was excavated from the northern Chilean Andes where she had died around 750 A.D. Her desiccated body was buried in a mummy pack along with her possessions: three woolen shirts, some feathers, corncobs, a wooden spoon, a gourd container, and a metal crucible. She was no Scythian queen. Her hair reached down her back in a long braid tied with a green cord. There were lesions in her spine, sternum, pelvis. On top of her skull, cancer had chomped a ragged hole 35 millimeters across. Cancer had feasted on her right femur, shortening her leg.

Osteolytic lesions are also found in men. They had spread throughout the skeleton of a Late Holocene hunter-gatherer exhumed in the

Argentine Pampas. Men do get breast cancer, but only very rarely. Lung cancer can also leave osteolytic marks, but it is believed to have been exceedingly uncommon before cigarettes. His diagnosis was left hanging. It was another case of what oncologists call "primary unknown."

Those words still haunt me when I think about the weeks that passed before finding the source of Nancy's metastasis. Like 90 percent of human cancers it was a carcinoma. It makes sense that these would be the most common. Carcinomas arise in the epithelial tissues that line the organs and cavities of the body and envelop us with skin. As the layers are worn by the passage of food and waste or exposure to the elements, the outer cells are constantly dying. The cells beneath must divide to form replacements. And with every division there will be mistakes in the copying of genes—spontaneous mutations or ones caused by carcinogens in food, water, and air. For children, who are just beginning to withstand life's wear and tear, only a fraction of cancers are carcinomas.

When it comes to hunting ancient cancer, primary carcinomas would almost always be lost with the decomposing tissues. And those that had metastasized would have often spread first to the lung or liver, killing the victim before a record was left in bone. Egyptian medical papyruses make ambiguous references to "swellings" and "eatings," and some evidence has survived in mummies. A rectal carcinoma in a 1,600-year-old mummy was confirmed with a cellular analysis of the tissue. Another mummy was diagnosed with bladder cancer. Elsewhere in the world, a rare muscle tumor called a rhabdomyosarcoma was found on the face of a Chilean child who lived between 300 and 600 A.D. In Peru, two pathologists reported metastatic melanoma in skin and bone tissue of nine pre-Columbian Incan mummies. In a whimsical digression, they quote an eighteenth-century ode in praise of female beauty marks and then wryly remark: "Whereas [the poet] was inflamed, as were his contemporaries, by the beauty of a lady's moles, we—some 240 prosaic years later—are romantically unmoved by any of them. They have given us nothing but trouble."

Other evidence of ancient cancer may have been destroyed by the invasive nature of Egyptian embalming rituals. To prepare a pharaoh for passage to the afterlife, the first step was removing most of his organs. The brain was pulled out though the nostrils. The torso was sliced open to take out the abdominal and chest organs (with the exception of the heart, which was believed necessary for the ethereal voyage). Each organ was wrapped in resin-soaked linen and then placed back into the body or in what was called a canopic jar. There were other variations. To slow the process of decay a turpentine-like solution was sometimes injected as an enema to dissolve the digestive tract.

But embalmed tumors can survive. Treated more gently, the mummified body of Ferrante I of Aragon, who died in 1494 in his early sixties, harbored an adenocarcinoma that had metastasized to the muscles of his small pelvis. Some five hundred years after his death, a molecular study revealed a typographical error in the DNA code that regulates cell division—a G had been flipped to A—a genetic mutation associated with colorectal cancer. Maybe this was caused, the authors speculated, by an abundance of red meat served in the royal court. Or, for all we know, by an errant cosmic ray.

Altogether I counted about two hundred suspected cancer sightings in the archaeological record. As with the dinosaurs, I was left to wonder how big an iceberg lay floating beneath the tip. Mummies are a curiosity, and most skeletal evidence is stumbled on by chance. Only recently have anthropologists really begun looking for cancer—with CT scans, x-rays, biochemical assays, and their own eyes. What they will never see, even in bone, are clues lost through what anthropologists call taphonomic changes. In digging and transporting skeletal remains, markings can inadvertently be erased. Bone-eating osteolytic lesions can cause a specimen to crumble and disappear. Through erosion, decomposition, and the gnawing of rodents, taphonomic changes might also create the illusion of metastasis—pseudopathology—a possibility that must be taken into account along with alternative diagnoses like osteoporosis and infectious disease. But on balance it seems likely that the evidence of ancient cancer is significantly underreported. Most skeletons, after all, are incomplete. Metastases are more likely to appear in certain bones like the vertebrae, pelvis, femur, and skull. Others rarely are affected. No one can know if a missing bone happened to be the one that was cancerous.

Hoping to cut through the uncertainty, Tony Waldron, a paleo-pathologist at University College London, tried to get a feel for how much cancer archaeologists should be expected to find. First he had to come up with an estimate, no matter how crude, of the frequency with which primary tumors might have occurred in earlier times. There wasn't much to go on. The oldest records that seemed at all reliable were from the registrar general of Britain for causes of death between the years 1901 to 1905. Using that as his baseline, he took into account the likelihood that various cancers would come to roost in the skeleton, where they might be identified. The numbers, a range of approximations, came from modern autopsy reports. For colorectal cancer the odds were low, 6 to 11 percent, as they were for stomach cancer, 2 to 18 percent. On the high side were cancer of the breast (57 to 73 percent) and prostate (57 to 84 percent).

From these and other considerations, Waldron calculated that (depending on age at death) the proportion of cancers in a collection of old bones would be between 0 and 2 percent for males and 4 and 7 percent for females. No matter how hard you looked, cases of ancient cancer would be sparse—even if the rate had been as high as that of industrial Britain. To test if his numbers were plausible, he tried them out on the remains of 623 people who had been placed in a crypt at Christ Church, Spitalfields in the East End of London between 1729 and 1857. Relying solely on visual inspection, he found one case of carcinoma among the women and none among the men. That was within the range of his formula, encouragement that it was not wildly wrong.

The next step was to try the predictions on much older and larger populations: 905 well-preserved skeletons buried at two sites in Egypt between 3200 and 500 B.C. and 2,547 skeletons that had been placed in a southern German ossuary between 1400 and 1800 A.D. (The church cemetery was so small and crowded that remains, once they had decomposed, were periodically removed and put into storage.) Using x-rays and CT scans to confirm the diagnoses, pathologists in Munich found five cancers in the Egyptian skeletons and thirteen in the German ones—about what Waldron's calculations predicted. For all the differences between life in ancient Egypt, Reformation Germany, and early twentieth-century Britain, the frequency of cancer appeared to be about the same.

Since then the world has grown more complex. Longevity has soared along with the manufacture of cigarettes. Diets have changed drastically and the world is awash with synthetic substances. The medical system has gotten better at detecting cancer. Epidemiologists are still trying to untangle all the threads. Yet running beneath the surface there is a core rate of cancer, the legacy of being multicellular creatures in an imperfect world. There is no compelling evidence that this baseline is much different now than it was in ancient times.

While still immersed in the arcana of paleo-oncology, I had dinner with a friend, a scientist in her thirties who had recently been treated for breast cancer. Like many people she suspected that there is far more cancer now than in the past, and a few weeks later she sent me a reference to an article that had just appeared in *Nature Reviews Cancer* in which two Egyptologists concluded that there is "a striking rarity of malignancies" in ancient times. In a news release from her university, one of the authors, A. Rosalie David, made this claim:

> In industrialised societies, cancer is second only to cardiovascular disease as a cause of death. But in ancient times, it was extremely rare. There is nothing in the natural environment that can cause cancer. So it has to be a man-made disease, down to pollution and changes to our diet and lifestyle.
>
> . . . We can make very clear statements on the cancer rates

in societies because we have a full overview. We have looked at millennia, not one hundred years, and have masses of data.

Across the Internet, news reports jumped on the information: "Cancer Is a Man-Made Disease." "Cure for Cancer: Live in Ancient Times." By now I thought I had become familiar with the literature. Was there some important new evidence that had resolved the ambiguities? It was flat out wrong to say that nothing in the natural environment can cause cancer. What about sunlight, radium, aflatoxin, hepatitis virus, human papillomavirus? I kept checking the university website assuming there would be a correction. None ever came.

The paper itself turned out to be more sober and qualified, and as I went through it line by line, I saw that nothing there was new. The authors had taken the same body of research I'd been wading through all winter and given it their own spin. While two hundred serendipitously documented cancer cases seem like a significant amount to most paleopathologists, some take the number at face value, envisioning an idyllic cancer-free past: a world where it was far less likely for children to get osteosarcoma or for even the very aged to get breast, prostate, or any of the cancers we worry about today. A world free from the attack of modern times. One can find consolation in fatalism, the idea that cancer is an inevitable part of the biological process. But there is also comfort in believing that humans, through their own devices, have increased the likelihood of cancer. What free-willed creatures have created can conceivably be undone. Failing that, there is at least a culprit to blame.

As I flashed back and forth between these opposing views, I was reminded of an optical illusion that at one moment looks like a beautiful young woman and the next moment like a crooked-nosed hag. With so little good data to go on, people see what they hope to see.

Seeking perspective, I wondered what fraction of the human bone pile had actually been picked. I asked three anthropologists to estimate the total number of ancient and prehistoric skeletons that have been discovered over the years and made available for study by the world's scientists. Perhaps 250,000, I was told, not much more than the population of a small city. That includes partial skeletons—and often only single skulls, which were the only bones many early anthropologists thought worth saving. Very few of the specimens have been scrutinized for cancer.

Take this number and compare it with the total number of people who have ever lived and died. A demographer at the Population Reference Bureau made a rough calculation. By A.D. 1, the earth's cumulative population had already approached 50 billion, and the number had nearly doubled to 100 billion by 1850. I was surprised by the magnitude. So much for the common notion that as many people are alive today as all who came before us.

Dividing 250,000 skeletons by 100 billion people you arrive at a few ten-thousandths of 1 percent. That is roughly the sample size on which our knowledge of ancient cancer is based—a sparsely dotted Rorschach you can choose to read two ways.

## Chapter 4

# Invasion of the Body Snatchers

On October 9, 1868, a patient identified in a style common to Russian novels and medical case reports as Richard J— was admitted to Melbourne Hospital with a diagnosis of "rheumatism and debility." He was weak, in other words, and his joints and muscles hurt. Almost anything could have been wrong. Beneath the skin of his chest and abdomen were about thirty lumps "varying in size from that of a bean to that of a small orange." There were two more tumors, one between his shoulder blades and the other on his inner left thigh about four inches above the knee. Over the next five months he wasted away, and after he died tissue from the tumors was prepared for examination under a microscope.

The physician in residence, Thomas Ramsden Ashworth, described what he saw: "large and beautifully pellucid cells" with distinctive features that left a deep impression on his mind. Because of the prevalence and the aggressiveness of the cancer, he was curious to see what the man's blood looked like, so he drew a sample. Floating among the red and white corpuscles, he was surprised to find cells that looked exactly like those inside the tumors. How did they get there? The blood sample had been drawn from a vein in the good leg, not the one that had been visibly affected by cancer.

The identity of the malignancy wasn't determined. An expert who examined the cancer had never seen one like it. More important to the history of medicine was the final observation in Ashworth's report: "The fact of cells identical with those of the cancer itself being seen in the blood may tend to throw some light upon the mode of origin of multiple tumours existing in the same person." He allowed for the possibility that the tumors might have formed spontaneously in the blood, either before or after death. Many physicians believed that cancer spread by secreting "morbid juices." But Ashworth suggested a more original hypothesis: that cancerous cells themselves had found their way into the bloodstream and transplanted themselves in distant locations. "One thing is certain, that if they came from an existing cancer structure, they must have passed through the greater part of the circulatory system." From the bad leg to the good leg, where they were ready to grow.

It was only in the nineteenth century that doctors had come to understand cancer as a disease involving abnormal cells. Hippocrates referred to "metastatic affections" traveling through the body. But he attributed cancer and other disorders to an imbalance of the body's four humors—blood, phlegm, yellow bile, and black bile, which were cosmically in tune with air, water, fire, and earth and with the primal qualities, hot, dry, wet, and cold. Those were the joints along which he carved up the world. If produced in excess, black bile (also called *melan cholo*) clotted to form tumors—an idea that was carried by Galen through the Middle Ages.

This conceptual stranglehold was loosened in the seventeenth century when René Descartes saw a connection between the recently discovered lymphatic system and cancer. This was a major advance— lymph, unlike black bile, was something that actually existed and could be observed—but there was still a long slog ahead. Veering off in the wrong direction, physicians began hypothesizing that tumors consisted of rotten lymph—not much of an advance over the notion of clotted *melan cholo*. A Parisian surgeon, Henry François le Dran, came closer to the modern view, proposing in 1757 that cancer began in a specific location—it was not some general malaise of the body—and then was transported in some form through the lymphatic channels, the blood, and sometimes into the lungs. The idea was slow to develop. Later on, metastases were thought to be transmitted by "irritations" traveling along the lymph vessel walls. Even the nervous system was said to be involved, dispatching signals to remote locations and causing the same kind of tumors to form. Comparing cancer with leprosy and elephantiasis, some scholars were certain that it also spread from body to body, that it was a contagious disease.

By the early nineteenth century physicians had noticed that "cancer juice" extracted from tumors consisted of tiny globular shapes. But the resolution of their microscopes was not sharp enough to show that they were actually observing biological cells. Helped by improvements in optical lenses, Johannes Müller, a German physiologist, made a crucial leap. In a book published in 1838, *On the Nature and Structural Characteristics of Cancer, and of Those Morbid Growths which May Be Confounded with It,* he laid out what came close to being a cellular theory of cancer. He saw with his microscope that a tumor consisted of cells, but he believed that they originated not from other cells but from a primitive fluid called blastema flowing throughout the body. Like his colleagues he couldn't shake the seductive image of tumors as some kind of clot.

Müller's student, Rudolf Virchow, took the next step, embracing the dictum *Omnis cellula e cellula*—all cells arise from other cells, including those that are cancerous. But when it came to explaining how cancer spread through the vessels he stumbled. He carefully considered the possibility that the process might involve "a dissemination of cells from the tumours themselves." But he found the notion of metastasis by "conveyance of juices" more plausible. Virchow also believed that all cancer arose from connective tissue, which we now know to be true only for the sarcomas, which account for a small portion of tumors. The German surgeon Karl Thiersch helped discredit that idea in the 1860s, showing that carcinoma arises from epithelial cells. Going further, he offered laboratory evidence that a tumor spreads by shedding its own cells, which migrate to other places. Thiersch is the source of one of the most depressing observations about cancer I have come across: "Cancer is incurable because it cannot be cured; the reason that we cannot cure it is because it is incurable; therefore, if by chance one should happen to cure it, it must be that there was no cancer."

As I tried to trace the flow of ideas that led to the modern theory, I was struck by how hard it is to tease out the subtleties of what any one person, no longer available for questioning, truly believed. It seems strange that doctors thought of cancer as a malign disposition of the whole body rather than a localized disease. But cancer would often be noticed only after it had broadcast itself far and wide. The idea of morbid juices sounds quaint and unenlightened, but there was a real question of how cancer cells, in their travels through the bloodstream, squeezed through the tiny capillaries of the lungs. The answer is still not entirely clear today. As always in science, people were playing with ideas, and more than one at a time. Streams of hypotheses emerged from hundreds of scientists as they engaged in slow-motion debate. The alternative to summarizing and schematizing and leaving names out is to take a plunge as deep as the German physician Jacob Wolff's. His tightly packed, elaborately detailed treatise, *The Science of Cancerous Disease from Earliest Times to the Present,* was published in four volumes beginning in 1907. They encompass 3,914 pages. An introduction to the first volume, the only one available in English, suggests that the reader "may or may not wish to compare [the work] with the magnitude of Pliny's *Natural History.*" Who knows what gems lie forgotten there?

By the time Thomas Ashworth saw what appeared to be circulating cancer cells, the modern theory of metastasis was falling into place. Next to be discovered was that these migrants would not take root just anywhere. After studying hundreds of cases of fatal breast cancer, Stephen Paget, an English surgeon, observed in 1889 that the malignancy usually traveled to the liver even though it could have just as easily reached the spleen. Metastasis was apparently not entirely a random event in which a cancer cell happened to become trapped by the narrows of a capillary or some other obstruction and then started to grow. It required the right environment. He was reminded of how plants replicate on the back of the wind. "When a plant goes to seed, its seeds are carried in all directions," he observed. "But they can only live and grow if they fall on congenial soil." This has become known as the seed and soil theory of metastasis: Different kinds of cancer seeds prefer different bodily tissues.

Despite Paget's insight, the belief persisted that it was nothing more mysterious than the layout of the vascular plumbing that determined where a cancer spread. The mechanics was clearly an important factor. There is a direct venous route from the colon to the liver, and the liver is the most frequent site of metastasis for colon cancer. Even if liver tissue didn't provide especially fertile conditions, it would soon be swamped by so many malignant cells that a few might chance to thrive. But other metastases are harder to explain. Bladder cancer cells often head straight for the brain.

As Paget's observations suggested, there had to be more to the process than proximity and dumb luck. In 1980 Ian Hart and Isaiah Fidler demonstrated this with a classic experiment involving laboratory mice. First they grafted fragments from different organs—kidney, ovary, and lung—beneath the animals' skin or within the muscle fibers and waited for capillaries to sprout, connecting the foreign tissue to the bloodstream. Once the grafts had taken hold, they injected the mice with melanoma cells that had been tagged with a radioisotope so that their paths could be followed through the body. While the malignant cells were equally likely to reach any of the three locations, cancers developed only in the lung and ovarian tissue.

A video I came across made these mysterious journeys seem a little less abstract. Beneath the microscope lens the edge of the tumor resembled a colony of tiny insects—the restless cancer cells. I knew I was watching a stochastic, mindless process, but it was impossible not to ascribe intentions and even feelings to the little devils. Some of them would venture timidly a short way from home. Startled by the foreignness, most quickly retreated to the safety of the pack. But occasionally a few particularly brave cells would crawl their way toward a blood vessel. The odds of making it very far were grim. When they become detached from their substrate, normal cells panic and initiate a preprogrammed suicide routine. The process is called anoikis, from the Greek word for "homeless." Some cancer cells apparently evolve the ability to overcome this fatal lonesomeness, but when they finally make it into a vessel, most will perish immediately in the river of blood—smashed against a vessel wall, pinched to death in an impassable strait, called out and destroyed by officious immune cells. So many dangers. I thought of the movie *Fantastic Voyage,* where a tiny team of doctors in a shrunken submarine faces one peril after another while exploring the human bloodstream. I thought of the great efforts experimental biologists take to keep cells alive in petri dishes. Some research suggests that the swimming cancer cells can surround themselves with a phalanx of platelets (blood-clotting cells) for protection during the journey. Or if stuck inside a capillary, some cancer cells may be able to jettison enough of their cytoplasm to slim down and squeeze through.

However they survive the journey they still must find a berth downstream. Here again most will perish. In other experiments with radioactively tagged cancer cells, researchers found that after twenty-four hours only 0.1 percent were still alive and that less than 0.01 percent went on to form tumors. The odds seem almost comforting, but of all the seeds a tumor can shed, it takes only one to start another cancer.

Cells are so particular about where they live that science still struggles to understand metastasis. How do the malignant cells decide where to go, and what counts for them as hospitable soil? Tissue similar to that found in the original tumor would surely be the most desirable, and yet cancer in one breast rarely moves on to the other breast. Nor does cancer in one kidney often spread to the opposite one. According to some theories, cancer cells wandering the corridors of the circulatory system are looking for a particular address—a molecular "zip code" identifying the organ where they are likely to thrive. Cancers are usually capable of replanting themselves, with varying success, in several kinds of tissue. In the Darwinian struggle inside a tumor different lineages may evolve specific genetic programs, priming them for survival inside the brain or, alternately, for a new life in the lungs. The primary tumor might smooth the way by secreting chemicals into the blood that help create a premetastatic niche downstream, a more hospitable place for the progeny to grow. There is even speculation that the travelers can bring their own soil with them—healthy cells from home that will assist in the colonization.

Once the cancer cells arrive at a promising location, a whole new cascade of events begins. They exchange signals with the natives—the cells of the tissue they are set to invade—recruiting their help in coming ashore. If cooperation is not forthcoming, the interlopers might lie dormant for years or decades until they are reawakened. When they have finally established their first colony, some will move on to other sites, and they may even return to the mother tumor to rejoin the battle at home. This self-seeding might help explain the recurrence of cancers that surgeons are confident they had completely excised. Metastasis—what would seem to be a messy, haphazard matter of tumors shedding cells willy-nilly into the bloodstream—turns out to be exquisitely and horrifyingly precise.


Besides the blood, there is another course the seeds can follow—from the tumor through the lymphatic vessels, making themselves known, as they did with Nancy, when they begin congregating inside a lymph node. I don't remember learning about the lymph system in school, this primitive, insect-like sewer system. It has no heart, sluggishly draining clear, watery waste from the cracks between cells, waste that is filtered along the way by the lymph nodes. Pushed and pulled by contracting muscles and osmotic pressures, the lymph eventually makes its way to the rushing blood, connecting with veins in the neck and shoulders. Evolution in its opportunistic way has found another use for the lymphatic canals: to transport immune cells called lymphocytes. These collect in the lymph nodes, rapidly mushrooming in number when confronted with foreign tissue— bacteria, viruses, cancer cells, enemies to destroy.

Malignant cells gain a pathway to the bloodstream when a tumor acquires the ability to initiate angiogenesis, growing its own capillaries. Tumors can also learn to induce lymphangiogenesis, creating connections to the lymphatic system. They may even send signals to a nearby lymph node, instructing it to sprout more vessels to accommodate the forthcoming invasion. The lymphatic system—this key component of the body's immunological defenses—becomes co-opted. The first sign is a tumor—a lump—growing inside a lymph node, the barrier whose very purpose is to stop such attacks. That apparently is what had happened with Nancy. It was why we were sitting, on what was probably a perfectly good autumn day, in an office at the university cancer center in Albuquerque.

For all the high-tech scanning and laboratory assays, the precise nature of her metastasis was confirmed by a procedure almost medieval in its barbarity: an endometrial curettage—scraping cells, in this case without an anesthetic, from the lining of the uterus for pathological scrutiny. To help endure the pain she was given a tongue depressor to clench between her teeth. After all the waiting, the procedure had to be done in a rush. We had been referred to a gynecological oncological surgeon, a specialist among specialists and a rising star in his field. He was leaving the next day for two weeks. To schedule surgery as quickly as possible the lab work had to be ready for his return. The results were what everyone by now had suspected: The cells from the uterus resembled those that had been found in the lymph node of her right groin.

On the scale of medical horrors learning that one has uterine cancer can be relatively good news (that is how far life had plunged).

Most cases by far are endometrioid adenocarcinomas—cancer of the epithelial cells of glandular tissue. Unlike ovarian cancer, it is usually noticed early and the five-year survival rate can be as high as 90 percent if the malignancy has not advanced beyond the uterine lining. If it has the odds are lower. When there is metastasis to the nearest lymph nodes (sentinel nodes, they are called, for they are the first line of defense against the errant cells) the likelihood of survival can drop to 45 percent—and if the cancer has advanced as far as an inguinal node, as it had with Nancy, to 15 percent. But those were just averages. Nancy's youth gave hope for a better than normal outcome. She was strong and could tolerate a regime of treatment—"regime" is precisely the right word—at least as aggressive as the cancer: multiple rounds of sickening chemo followed by burning radiation. But first would come the surgery. A hysterectomy, of course, and removal ("dissection") of suspect lymph nodes. The surgery would also be exploratory with the aim of identifying and excising any other tissues the cancer might have invaded.

The operation was scheduled for early November, still weeks away. All that time to imagine the cells as they continued to multiply, trying out new combinations of mutations. We went to a lawyer to draw up living wills and medical powers of attorney. Nancy's youngest brother flew in from the East Coast to be with us. One night shortly before surgery we were sitting together in a Thai restaurant (it's strange the details one remembers) pretending to be enjoying dinner. During the meal Nancy mentioned that she had noticed a lump that day in the inguinal node of her left leg. The good one. Remembering this now I think of that 1868 paper by Thomas Ashworth: One thing was certain. Moving through her lymphatic system, the cancer cells had reached the other side of her body. And they had found hospitable soil.

As I learned about metastasis, I thought about the years before the cancer when Nancy and I worked so hard to turn a desiccated, junk-strewn weed patch—our backyard—into a xeriscape garden. Not a zeroscape—those gravel and cactus afterthoughts one sees in Phoenix or Las Vegas—but something akin to a dry highland meadow. We started with one small patch, clearing it of brush and scattering a packet of Beauty Beyond Belief wildflower seeds, a mix recommended for northern New Mexico. There were seeds for Colorado aster, goldfields, arroyo lupine, desert lupine, desert marigold, California poppy, alyssum, baby blue eyes, baby's breath, bachelor button, black-eyed Susan, candytuft, catchfly, columbine, purple coneflower, yellow coneflower, coreopsis, cosmos, African daisy, Shasta daisy, blue flax, scarlet flax, mountain garland, gaillardia, larkspur, perennial lupine, Mexican hat, Rocky Mountain penstemon, corn poppy, sweet william pinks, and wallflower. We raked them into the dirt and let nature take its course.

When the rains came it was clear that all we were going to get was blue flax, coneflower, and Mexican hat. They overflowed the garden and over the years found niches throughout our irregularly shaped quarter acre of land. The yellow coneflower and Mexican hat, both members of the genus *Ratibida,* mated to form hybrids that still appear each season. On Saturday mornings we would come home from the nursery with flats of new wildflowers to try. For all our efforts some would die not long after planting, but those that survived would set seed in the fall. The winds would come, then the rain, and we would find Rocky Mountain penstemon and red pineleaf penstemon in surprising new places. They would grow there and thrive in a way they never did when we were the ones to choose the locale.

Some wildflowers native to the foothills where we lived flourished along the trails. Yet they were nearly impossible to cultivate: *Hymenoxys argentea* with its silvery leaves and yellow flowers, *Phlox nana* (locally called Santa Fe phlox), which bloomed little violet stars. A local nursery managed to grow only a few of the plants and there was a waiting list each spring. It took years of trial and error until the phlox finally found a spot, shaded by a pine tree, where it deigned to grow. Nancy had majored in biology and would show me how a leaf of a wildflower began to change at the tip, gradually in shape and color, until one day there was a blossom. It had never occurred to me that the same green cells that formed the leaf were differentiating into colorful petals—genes switched on and off, signaled by sunlight, temperature, moisture, whatever told the plant that it was time to bloom. Differentiation and development could occur at astonishing speeds.

What adapted far more readily were the weeds. After our first summer rain in Santa Fe, a bluish green carpet that we welcomed as some unidentified native ground cover turned out to be seedlings of kochia, a member of the goosefoot family that originated in the harsh climate of the Russian steppes. For all its aridity, New Mexico must seem to this immigrant like a tropical paradise. The tiny plants rapidly shot up to form ugly, spindly weeds.

Another hated intruder from Eurasia was western salsify, and we thought at first that it was no worse than a larger version of the American dandelion. We quickly learned better. One morning we were showing our fledgling gardens to our neighbor Vivian when she spotted one of these weeds, now more than a foot tall, with a podlike bud protruding outward that was about to open into a flower. Vivian shrieked melodramatically and pulled it up by its roots, advising us to kill every one we found. As we soon learned, the pretty yellow petals would turn, seemingly overnight, into a cloud of feathery white seeds, each so viable that western salsify would quickly spread throughout the yard outcompeting almost everything. It spread so viciously that we imagined it, in the dark of night, expectorating its deadly spores in one explosive burst. We thought of the pods in *Invasion of the Body Snatchers,* landing from some distant star to take over the earth. We nicknamed the weed "space plant," and I learned to recognize and destroy the seedlings when they were barely half an inch high.

That was a few years before Vivian died of ovarian cancer. The spreading of weeds became linked in my mind with metastasis. But maybe that was the wrong metaphor. Cancer, as Paget realized so long ago, is more discriminating in the way it propagates. Honed for life in a specific tissue, a metastasizing cancer cell had more in common with those delicate wildflowers—until it found its roost. Then it was more like the pods.

## Chapter 5

# Information Sickness

The first hint that cancer is a disease of information came in a laboratory at the University of Texas, where in the late 1920s Hermann J. Muller was experimenting with fruit flies. He was working in a long tradition that had begun with Mendel, who discovered in his monastery garden that certain traits like flower color are passed down among generations of pea plants according to predictable patterns. Purpleness is a dominant factor and whiteness is a recessive one. If a pea plant inherits the purple factor from each parent, its flowers will be purple. The same rule holds true if both inherited factors are white. But if one is white and the other is purple, they do not blend to make lavender. Purple trumps white so that is the color that appears in the progeny. The modern way of saying this is that there is a gene for flower color—a microscopic kernel of hereditary information—and that it comes in two forms. With fruit flies, which breed so rapidly, the shuffling of these tokens unfolds in fast-forward. Eyes red or white, bristles straight or forked—these genetic traits, as discrete as the ones and zeroes of binary code, can be followed and plotted as they travel down the family line.

As a student, Muller had studied how the Mendelian process

sometimes spits out a wild card. After many generations, purebred red-eyed flies would spontaneously produce a mutant with white eyes. Other kinds of mutations would also appear. This was long before DNA was identified as the stuff of genes, the helically shaped molecule that carries genetic information in a four-symbol alphabet— the nucleotides abbreviated G, C, A, and T. If a letter is changed, the meaning can be corrupted. The signal becomes noise or is silenced altogether. That kind of clarity would come decades later with the discoveries of Oswald Avery in 1944, Alfred Hershey and Martha Chase in 1952, and a year later when James Watson and Francis Crick cobbled together from cardboard, sheet metal, and wire their model of the double helix. For now Muller's contribution was to show that whatever genes were made of and however they worked, you didn't have to wait for mutations to occur. They could be produced at will by exposing the flies to x-rays.

Most often the mutations sterilized the flies or killed them. That, he speculated, might explain why the rays were so effective at destroying rapidly dividing cancer cells—a therapy that had come into use almost as soon as x-rays were first produced in the laboratory of Wilhelm Röntgen in 1895. With each cellular division the genes had to be copied. The energy from a penetrating x-ray could damage the microscopic structure, inducing a lethal mutation and removing the cell from the game. Far more telling was that Muller's x-rays could also create living mutants: albino fruit flies or fruit flies with forked bristles or shrunken wings. This ability to alter genetic material, he suggested, might explain a paradox: why the cancer-killing rays could also *produce* cancers, transforming normal cells into malignant ones. Cancer, this seemingly amorphous disease, this sprawl of hyped-up cells, might be the result of precise genetic mutations.

The clues had been lingering barely visible since the early 1900s when a German biologist, Theodor Boveri, wondered why cancer cells had strange-looking chromosomes. Maybe, he speculated, they were damaged in a way that knocked out "factors," whatever they might be, that would normally rein in growth, allowing the cells to "multiply without restraint."

Reverting to a more primitive state, a cancer cell abandoned its communal obligation to replicate only when "the needs of the whole organism require it." What had been a responsible member of an organization became like a single-minded paramecium whose only aim, Boveri wrote, is to egotistically propagate itself. Half a century before DNA was decoded, he even ventured that a cancer cell goes native because "chemical and physical interventions" damage some of its internal workings without killing the cell outright. He was writing in 1914. Five years later, inspired by Boveri, the geneticists Thomas Hunt Morgan and Calvin B. Bridges found it "conceivable at least that mammalian cancer may be due to recurrent somatic mutation of some gene." Another scientist spoke of cancer as "a new kind of cell" in which "an ever recurring process of mutation is taking place, with a tendency, however, to deviate more and more from the normal type." It is as impressive as it is frustrating how close they came to the mark.

Evidence had also been accumulating that radioactivity, like x-rays, was capable of causing mutations. Since ancient Rome, uranium had been mined and extracted from a rock called pitchblende for use as a yellow pigment in making glass and ceramics. No one knew of its more exotic qualities until 1896 when Henri Becquerel accidentally discovered that uranium salts wrapped in opaque paper or shielded with aluminum would fog photographic plates. He thought at first that the crystals were absorbing sunlight and then reemitting these piercing rays. What a chill he must have felt when he realized that the uranium was not sucking up the energy but producing it—this invisible and piercing light.

The situation only grew stranger when Marie Curie noticed that pitchblende retained its power even after the uranium was removed—in fact, the leftover ore was far more radioactive than the purified uranium itself. There must be something else in the rock that was even hotter. She and her husband, Pierre, isolated and named a new radioactive element, polonium (after Poland, her native land), only to find that the rock remaining was still extremely radioactive. Something still hidden inside was shooting out these incredible rays.

"Pierre, what if there is a kind of matter in the world that we've never even dreamed of. . . . What if there exists a matter that is not inert but alive?" That's Greer Garson, playing Curie in the 1943 movie *Madame Curie,* in a scene as erudite as it is melodramatic. In a drafty shed at the University of Paris she sifts through piles of pitchblende and extracts the tiniest speck of what she names radium. In the best part of the movie she and Pierre come upon the shed at night and find it shining with an eerie glow. The real story, uncompressed and undramatized, is just as moving. Here is how Curie described it in her own writings: "One of our joys was to go into our workroom at night; we then perceived on all sides the feebly luminous silhouettes of the bottles or capsules containing our products. It was really a lovely sight and one always new to us. The glowing tubes looked like faint, fairy lights." What the Curies were witnessing were contrails of light produced by charged particles shooting through the air, an optical analog of a sonic boom.

Radium also glows when its rays strike a phosphorescent chemical like zinc sulfide, and before long the two substances were mixed to produce glow-in-the-dark watch dials. Painting the numbers was a painstaking task—the hook at the top of a 2 thinning just so to produce the narrow downstroke, thickening again to form the base line. The numerals 3, 6, and 8 were equally demanding. To clean the tips of the brushes and keep them pointed, workers were trained to wet and shape them with their lips and tongues. Assuming that the paint was harmless, some of the dial painters—they became known in news reports as the Radium Girls—used it to decorate their teeth, fingernails, and eyebrows. It must have been great for Halloween.

Mistaken by the body for calcium, radium became incorporated into their bones, where it sat firing off high-speed electrons, alpha particles, and gamma rays, killing cells or transforming them and eventually giving some of the women cancer. Here was the paradox again: Curie herself had been promoting radium, like x-rays, as a therapy for shrinking cancerous tumors. But here it was producing tumors from healthy cells. In 1927 when the Radium Girls were making headlines, Muller's paper appeared, speculating that the mutagenic power of x-rays might be responsible for their ability to cause cancer. If so, then the same was probably true for radium's fairy light.

Long before invisible rays became a suspect, doctors were seeing clues that cancer could also be caused by more tangible stuff. In 1775 a London surgeon realized that "soot warts," sores appearing on the scrotums of chimney sweeps, were not venereal disease but a malignancy—apparently caused when skin came into contact with the black tars and dust left by burnt coal. The same cancer was later found in workers who manufactured paraffin and other coal tar distillates, and by the early twentieth century scientists were producing carcinomas by repeatedly applying coal tar to rabbits' ears. Coal tar was found to consist of a witch's brew of carbon-based compounds—benzene, aniline, naphthalene, phenols—and during the next few decades scientists discovered that many of them produced tumors in laboratory animals. It would have been unethical for them to expose human subjects to the carcinogens to see if they caused cancer. They didn't have to. With the growth of the cigarette industry, people were performing the experiment on themselves.

By the time the century was half through we knew that radiation caused both mutations and cancer. We knew that a host of different chemicals also caused cancer, and many of these were soon shown to be mutagens. They altered a cell's genetic software by changing snippets of the DNA code. In the early 1970s Bruce Ames (the scientist best known for showing that ordinary fruits and vegetables contain carcinogens) came up with a striking demonstration. Instead of fruit flies, he worked with salmonella bacteria—strains that had lost the recipe for making histidine, an amino acid they needed in order to reproduce. If placed in a dish of nutrients with a dash of this vital ingredient, the bacteria would grow, but only until they had depleted the supply. Then the whole colony would die. Ames discovered that if carcinogens were added to the mix, some of the salmonella would keep on living, expanding and overtaking the dish. The chemicals were presumably producing mutations at random. But each bacterium's genome carried so little information, and there were so many of the microbes—billions of them—that the mutations would include ones that happened to restore the ability to synthesize histidine.

The procedure came to be called the Ames test—a fast and dirty way to see if a chemical might be mutagenic. In instance after instance, chemicals that passed the Ames test also produced tumors in laboratory animals. The case almost seemed clinched. What causes cancer, whether chemical or energetic, does so by altering genetic information. The pieces of a theory were falling into place, except for a stubborn exception—at least some cancers appeared to be caused neither by chemicals nor penetrating rays but by viruses.

In retrospect that is not so surprising. Existing on the boundary between chemistry and life, viruses are packets of information—streamlined sequences of DNA or RNA wrapped in a protective sheath. They are wandering genomes so simple that some consist of only three genes. Like the handmade Internet viruses they later inspired, they infiltrate their hosts (the biological computers called cells) and commandeer the internal machinery. There the invader's genes are dutifully duplicated and repackaged again and again, the viral copies spreading to other cells where they robotically carry out the same routine—life itself stripped of its capacity to do anything except reproduce.

A few viruses operate in an even more convoluted way. They copy and splice their genes directly into a cell's chromosomes. This infiltrating algorithm orders the host itself to replicate at an accelerated pace. It becomes a cancer cell. The earliest example was reported in 1910 by Peyton Rous, a scientist at the Rockefeller Institute for Medical Research who was studying chicken tumors. He began by extracting fluid from an irregularly shaped glob growing in the breast of a Plymouth Rock hen and then injecting it into another bird. Thirty-five days later the first chicken had died from the cancer, a sarcoma, and the second chicken had developed a tumor of the same kind. Material taken from the tumor could, in turn, be used to spread the cancer to another bird. And so it went from fowl to fowl. The transforming agent turned out to be a retrovirus—the kind that can smuggle cancer-causing genes into otherwise healthy cells.

There was *src,* which was part of the virus that caused sarcoma in chickens. Another gene, called *ras,* induced sarcoma in rats, while *fes* did the same in felines. *Myc* and *myb* induced blood cell cancers, myelocytomatosis and myeloblastosis, in poultry. If that is where the research had ended it would have made for a tidy picture. Cancer could be caused when chemicals or radiation mutated preexisting genes, or when viruses surreptitiously inserted entirely new ones— oncogenes, they were called—already capable of causing cancer. Two fundamental ways of modifying genetic information. But the real story turned out to be far more interesting.

There was a problem reconciling Rous's discovery with what appeared to be happening in the world. Cancer wasn't acting like a contagious disease sweeping through populations like polio. It arose sporadically in various places. Even Rous's chicken virus spread only when it was injected, and try as he might he couldn't transfer it to other animals—pigeons, ducks, rats, mice, guinea pigs, rabbits. Only with great difficulty could it be induced in other chickens except closely related Plymouth hens. Even more suggestive, scientists were not finding the retroviruses inside human tumors. What they were discovering instead was that genomes of creatures throughout the animal kingdom contained what appeared to be naturally occurring versions of *src, ras, fes, myb, myc*—not ones that had been smuggled in. These were not broken, mutated genes like their viral counterparts. Their purpose was to govern how healthy cells divide, the process biologists call mitosis.

This is apparently what was happening: Occasionally a virus going about its rounds would accidentally copy one of these innocent "host" genes into its own genome. Passed along from virus to virus, the gene mutated into a form that caused cancer. But all that was a fluke. The virus was an accidental player in the story, the place where the first of these genes happened to have been discovered. Some cancers can be directly caused by a viral invasion—human papillomavirus and cervical cancer, hepatitis viruses and liver cancer. But these are exceptions. Far more often cancer arose when the original gene, sitting secure in its own cell, underwent a random mutation, one caused externally by a carcinogen or internally by an unprovoked copying error. One way or another the gene's normal function became distorted, tipping the cell toward malignancy. Since genes like these were capable of transmogrifying into a cancer gene, they were named proto-oncogenes. Had their true function been discovered before their aberrant one, they would be called something else.

Studying the genes more closely, researchers discovered how they regulate the ways in which cells grow and multiply in harmony. Some of the genes controlled the production of receptors that protruded from a cell's surface—molecules tuned to respond to signals from other cells. When these molecular antennae received a message they would relay the information internally to their own cell's nucleus—instructions to activate the machinery for dividing into daughter cells. If the gene became mutated the cell might produce too many receptors or overly sensitive ones. Spooked into responding to silence, they would pummel the cell with false alarms. Still other broken genes might unleash messages urging the cell's neighbors to flood it with more growth-stimulating chemicals. Or, in its hyped-up state, the cancer cell might overreact to its own signals, screaming at itself to grow.

Genes related to *src* are mutated in colon and many other cancers. Crippled *ras* genes show up in a variety of human malignancies—pancreatic, colorectal, thyroid, melanoma, lung. All that it takes to turn a good *ras* into a bad *ras* is a single point mutation—a G flipped to T, A, or C—a random typo in a message hundreds of letters long.

Other mutations occur during cellular division when a normal gene is copied too many times. Repeating *ras* genes are found in lung, ovarian, bladder, and other cancers. Stuttering *myc*s help give rise to a childhood brain cancer called neuroblastoma. Some mutations are even more wrenching: A chromosome might break and then join with another, placing two previously distant genes side by side. In Burkitt's lymphoma, a mutation like this shoves a *myc* gene next to an overbearing stranger that drives its new partner to overexpress itself, churning out signals that cause the cell to divide and divide and divide.

It was a terrifying possibility—that a single mutation might be enough to shift a gene into overdrive and give rise to a deadly tumor. But not even an oncogene wields so much power. Researchers found that inserting one or even two oncogenes into a cell was usually not enough to ignite a cancer—unless the cell had already accumulated some earlier defects. Living systems are governed by a gyroscopic balance in which an extreme force from one direction is met with a countervailing shove. While the 1970s was the decade of the oncogene, in the 1980s scientists began discovering anti-oncogenes—genes whose purpose was to respond to rapid bursts of cellular division by slowing the process down.

Like the proto-oncogenes, these growth-restraining genes were part of a cell's normal regalia, and they too were discovered when something went wrong. Retinoblastoma is a childhood cancer marked by runaway growth of the light-sensing cells of the eyes. The first sign might be an eerie white glow in the gaze of a child photographed with a camera flash. If noticed early enough the condition can be treated with chemo, radiation, laser surgery, or removal of the eye. If not the outcome can be horrifying, the expanding tumor expelling the eye from its socket. Pictures in nineteenth-century textbooks show the gruesome results, which still occur among the poor in developing countries. The cancer begins when a gene called *Rb*, short for "retinoblastoma," has been taken down by a mutation, losing its ability to curb excessive growth.

But *Rb*, named like so many others because of the accidental circumstances of its discovery, didn't exist for the sole purpose of suppressing retinoblastoma. Once scientists started looking for *Rb* genes, they found them throughout the body—and they were missing or crippled in cancers of the bladder, breast, and lung. Unlike an oncogene like *myc* or *ras,* growth-restraining genes like *Rb* are conspicuous by their absence. Because we inherit the chromosomes of both parents, genes exist in pairs. In a single cell, only one oncogene needs to start misbehaving for trouble to begin. With genes like *Rb,* both copies must be knocked out. If only one is lost the other will still be there to send moderating signals.

Dozens of similarly purposed genes have been discovered: *PTEN, apc, vhl, p53*—"tumor suppressors," another awkward name thrust on the world by the human tendency to notice things only when they break. In an old-fashioned radio one can reach in with gloves and remove a hot glowing vacuum tube from its socket, unleashing a blasting squeal from the loudspeaker. Someone coming upon the phenomenon for the first time might name the component a squeal suppressor. But the circuitry is so much more complex. So it is with the suppressor genes. Some produce receptors that listen for inhibitory signals—orders from neighbors to stop overstepping their bounds. Others code for enzymes that muffle the commands of growth-stimulating genes. The rhythm of cellular division is governed by the molecular gears of a cell-cycle clock, and tumor suppressor genes are also involved in the timekeeping.

One of them, *p53,* sits at the center of a web of chemical pathways controlling the life cycle of a cell. If you want to start a cancer, take down *p53.* If a cell is damaged and dividing too quickly, external sensors will pick up warning signals from crowded neighbors. Internal sensors will detect chemical imbalances or broken DNA. With an emergency declared, *p53* will step in and slow down the clock so that DNA repair can take place. Proofreading enzymes scan the genome. If one strand of DNA's double helix has been corrupted, the other strand can be used as a template to guide repair. Damaged sections can be excised, a replacement synthesized and put into place.

If DNA repair is broken and other measures cannot save a cell that is mutating beyond control, *p53* initiates programmed cell death, or apoptosis. The name is derived from a Greek word describing falling leaves. When an embryo is developing into a little body, it will produce far more cells than it needs. Apoptosis is the means by which it sheds the excess. Webs between fingers and toes are pared back. Lumps of neurons are sculpted into a thinking brain. Apoptosis is not just one big cellular explosion but an intricate procedure in which death signals set off the molecular equivalent of strategically placed depth charges. The nucleus implodes, the cell's cytoskeleton crumbles. The microscopic remains are engulfed by other cells and a would-be malignancy is gone.

Through random mutations a few cells will learn to thwart or ignore the death signals—and then double and double and double again. A normal cell can divide only fifty or sixty times—a principle called the Hayflick limit. The count is kept by telomeres, caps on the ends of chromosomes that get a little shorter each time around. Once the telomeres fall below a certain size, mitosis comes to a halt and the worn-out cell is taken offline. Cells like those in the immune system, which must divide repeatedly, manufacture telomerase, an enzyme that keeps putting the caps back on the ends of the chromosomes. Cancer cells have also learned this trick, acquiring through the trial and error of mutation the information needed to produce their own telomerase. They can replicate indefinitely.

Conferred with the closest that nature has come to immortality, the cell and its descendants increase exponentially in number, each division giving rise to a new branch of the family tree. The branches divide fractal-like into more branches, and each of these lineages— these many-forking paths—is accumulating mutations. Equipped with different routines and survival skills, the clans compete for dominance.

As this evolution unfolds, the tumor that is emerging acquires more of the tools of carcinogenesis. Enzymes called proteases eat into healthy tissue. Cell adhesion molecules hold the expanding mass in place. Taking the invasion to a whole new level, signals are sent to healthy cells recruiting them to join the attack. Cells called fibroblasts obediently synthesize proteins for the tumor's structural support. Endothelial cells—those that line the circulatory and lymphatic systems—are summoned to help make the vessels that nourish the tumor and provide avenues for metastasis. Macrophages and other inflammatory cells, flocking to fight the invasion, are persuaded instead to aid in its expansion—producing substances that stimulate angiogenesis, lymphangiogenesis, and the creation of more malignant tissue. Here lies another paradox of cancer. The panoply of devices normally employed to heal a wound—destroying old diseased tissue and replacing it with healthy new growth—is turned on its head, subverted to promote malignancy.

All of these mechanisms are so intertwined that it can be difficult to tell where one leaves off and another begins. What is being done by the cancer cells and what is being done by its minions? Tumors were once thought of as homogeneous clumps of malignant cells. Now they are compared to bodily organs—systems of interlocking parts. There is a crucial difference. Organs are linked into a network of other organs, each playing an established role. A tumor is attempting to become independent, as though a kidney had decided to break free and set out on a life of its own.

Chapter 6

# "How Heart Cells Embrace Their Fate"

In a very creepy way, an embryo is so much like a tumor that the early days of pregnancy resemble the incursion of a malignant growth. Once an egg is fertilized, it travels down the fallopian tube, dividing and dividing along the way. After several days it has become a ball of dozens of identical cells, which proceed to gather themselves into two regions. The outer layer will become the placenta, while the inner cell mass will give rise to the fetus.

Exchanging signals with the uterine wall, this expanding mass, called the blastocyst, prepares to implant itself, the next step in a successful pregnancy. To carve out an opening, protein-dissolving enzymes erode the surface of the uterine lining. As the blastocyst digs in, a process embryologists call invasion, cell adhesion molecules help ensure a tight grip. Normally such an interloper would be rejected as foreign tissue, but messages are sent to the immune system enlisting its cooperation. If all goes as planned, blastocyst becomes embryo, and it begins stimulating angiogenesis—growing vessels to hook into the mother's blood supply. Every step of the way the molecular interactions of pregnancy are like those that occur during the genesis of a tumor.

As the occupation continues, the cells inside the fetus begin spreading in a well-orchestrated metastasis. First they gather themselves into three layers—the endoderm, mesoderm, and ectoderm (inner, middle, and outer). Cells from each of these primordial regions then strike out on their own, moving into new positions. As they travel they begin to differentiate. Bone and cartilage go here, dermis goes there, with nerves and blood vessels strung through. What began as identical totipotent stem cells—blank slates—become the specific cells of the body. There is no central overseer. Every cell contains the entire genome, and as the diaspora continues genes are turned on or off in different combinations, producing the unique set of proteins that gives a cell its identity. The endodermal cells give rise to the lining of the digestive and respiratory tracts and form the liver, gallbladder, and pancreas. The cells of the mesoderm form muscles, cartilage, bones, spleen, veins, arteries, blood, and heart. The cells of the ectoderm form skin, hair, and nails and also the neural crest, which develops into the nervous system and brain.

While tumors evolve through random mutations, fetuses do so according to a plan. But the deeper biologists look, the more parallels they find. As the fetus develops, tightly connected epithelial cells—the kind that form tissues—must loosen their grip so they can move to new locations. They become wanderers called mesenchymal cells. When they reach their destination they can turn back into epithelial cells and regroup into new tissues. This process, called the epithelial-mesenchymal transition, or EMT, also occurs during healing, when cells are dispatched to repair wounds at distant sites. It seems only natural that cancer would find a way to adopt EMT as a vehicle of metastasis, and there is compelling evidence that it does. Carcinomas, the most common cancers, are derived from epithelial cells. By temporarily changing identity they could more easily disperse through the body. During the transition they might even acquire properties like those of fetal stem cells—the ability to replicate profusely and generate a new tumor. There would be no need for the cancer cell to stumble upon these chameleon talents through random mutations. The program, left over from early days, would be waiting ready-made in the genome like a book forgotten on a shelf. It would simply have to be reread.

Wanting to learn more about the complex processes of life and anti-life, I drove down to Albuquerque one morning where the Society for Developmental Biology was holding its annual meeting. The essence of the science is to tamper with genes that play a role in embryonic unfolding and then see what kind of deformities occur. Experimenting with insects, worms, fish, and other laboratory creatures, biologists are slowly piecing together the steps that lead from a fertilized egg to a fully formed adult. Like ants in amber, the same cellular processes have been preserved and carried through evolution's forking paths. When activated at the wrong time they can bring on human cancer.

There was a flood of new results to impart since the previous year's meeting. The only way to accommodate them all was by running sessions simultaneously. "Organogenesis," "Spatiotemporal Control in Development," "Branching and Migration," "Generation of Asymmetry"—a feast of strange, enticing ideas. Darting from one room to another, I could sample the latest reports about the genes directing the development of the liver in the zebrafish or the brain of the sea squirt, or those that ensure that the trachea properly separates from the digestive tract in the embryonic mouse. One could learn about how sex is determined in the worm *C. elegans,* or how apoptosis—programmed cell death—sculpts the genitals of fruit flies. There were talks about how amphibians and planaria regenerate amputated body parts—and speculation about why that is something mammals cannot do.

Many of the genes directing development were first discovered in fruit flies. When mutated or destroyed, they cause deformations, and so they have been given names like wingless, frizzled, smoothened, patched, and disheveled. Mutations in a gene called hedgehog can cause bristles to grow unexpectedly on the undersides of fruit fly larvae. (A human hedgehog gene is involved in the sprouting of hair from follicles, suggesting possible treatments for baldness.) Genes called snail, slug, and twist are invoked in the gyrations of the epithelial-mesenchymal transition.

As scientists discovered variations, they went even wilder with the nomenclature. Desert hedgehog, Indian hedgehog, and sonic hedgehog. A gene called fringe was soon joined by manic fringe, radical fringe, and lunatic fringe. When mutated during the formation of an embryo the result can be deformities and neonatal cancer. The silly nomenclature has caused some uneasiness among people dealing with the heartbreaking outcome of developmental defects. One medical researcher put it like this: "The quirky sense of humour . . . often loses much in translation when people facing serious illness or disability are told that they or their child have a mutation in a gene such as Sonic hedgehog, Slug or Pokémon." The latter, proposed as a name for an oncogene, was withdrawn after the threat of a lawsuit from Nintendo, the maker of the Pokémon game. It now goes by the less evocative name Zbtb7.

No lawsuits were filed by Sega when its video game character, Sonic the Hedgehog, was appropriated by biologists. Even if the company had been inclined to sue, it was soon too late. Since it was discovered in 1993, sonic hedgehog has rapidly emerged as one of the most powerful components of animal development. The first hints came in the 1950s when sheep grazing in the mountains of Idaho were giving birth to deformed lambs. In the most hideous cases, there was a single eye in the center of the forehead, and oftentimes the brain had not completely divided into left and right hemispheres. After spending three summers communing with the sheep, a Department of Agriculture scientist discovered the cause. Drought was prodding them to wander higher up the mountains, where they dined on a lily called *Veratrum californicum*. Laboratory experiments confirmed that pregnant sheep that ate the plant gave birth to cyclopean mutants. The mutagenic chemical was isolated and named cyclopamine. It worked, biologists went on to discover, by suppressing the signals of the sonic hedgehog gene. (Sheep also played a part in the episode of the *Odyssey* where Odysseus and his men visit the island of the Cyclopes. Trapped in a cave they are devoured, one by one, by the one-eyed monster Polyphemus, until Odysseus blinds him with a homemade spear. He and his soldiers escape by tying themselves to the underside of Polyphemus's flock.)

In one session after another in Albuquerque, sonic hedgehog was there. It sets in motion a complex molecular cascade—what biologists call the shh signaling pathway—that also involves patched, smoothened, and other genes. In mammals, sonic hedgehog helps establish the left-right symmetry of the body and brain and guides the patterning of the skeleton and nervous system, linking bones with muscle and clothing them with skin. A dose of cyclopamine is not the only way to gum up the works. In the developing embryo, mutations can suppress sonic hedgehog, giving rise to a human deformity called holoprosencephaly. As with the lambs, the baby's brain doesn't properly bisect into hemispheres. There may be a nose with a single nostril or a mouth with one instead of two front teeth and, in the most severe cases, a cyclopean eye centered like a headlamp in the face. So many things have to go right during the formation of a child—the proper chemical signals produced, transmitted, and received at the proper locations, in the proper concentrations, at the proper times. More often than we realize, something goes wrong. It has been estimated that as many as one of every 250 early embryos is holoprosencephalic. These pregnancies usually end in a miscarriage, so the defect appears in only about 1 in 16,000 live births. Most of these babies die, but those with milder symptoms can live for years.

While too little sonic hedgehog signaling can cause birth defects, too much can drive the formation of malignancies both in children and adults: a brain tumor called medulloblastoma, for example, and basal cell carcinoma, the most common (and usually innocuous) form of human cancer. These skin growths tend to arise slowly and are easily excised in a dermatologist's office. But in people with Gorlin syndrome, hyperactive hedgehog behavior can cause hundreds of the carcinomas to appear. One study found that a cream containing cyclopamine beat back the growths, and a treatment involving another hedgehog inhibitor has been approved by the Food and Drug Administration.

The morning talks left me feeling frazzled (that is also the name of a gene, and sizzled is too), and I decided to take a quiet stroll through the poster session. In what has become a tradition at scientific meetings, row upon row of corkboards were arranged so scientists—usually graduate students and freshly minted PhDs—could pin up large placards describing in pictures and words some of their experimental achievements. Years ago when I was haunting neuroscience conferences, grazing the posters helped me acquire a lay of the land. Again I found myself becoming immersed in exciting, sometimes bewildering new territory. On this particular afternoon there were 148 posters on developmental biology, and many of the researchers were standing ready to run through the details.

Heading down one of the aisles, trying to avoid being buttonholed, I lingered for a moment at a seemingly unmanned presentation titled "Novel Transcription Factor Involved in Neurogenesis."

"Would you like me to explain my poster?" A young woman had suddenly appeared. I saw on her name tag that she was Emma Farley from Imperial College in London. I usually preferred to struggle through the posters in solitude but her enthusiasm was hard to resist. Starting at the upper left-hand corner, she explained how a molecule, Dmrt5, equipped with a molecular digit called a zinc finger, might help control the genetic switches during the maturation of the brain. The experiments were with mice and chickens. I followed as best I could as she periodically glanced at my face for signs of comprehension. At what level should she calibrate her explanation?

"What animal do you work on?" she finally asked. *Drosophila, Xenopus, C. elegans . . .* so many possibilities. I told her I was a science writer. She ratcheted down the level a couple of notches until I got the gist. Grateful for her patience, I walked to the lobby, sat down with my laptop, and googled "zinc fingers," "Dmrt5," and "Emma Farley," seeing that she had received a prize for an earlier version of her poster. Piece by piece I was putting together a map.

Once you stumble upon a strange new word, your brain seems

to sprout receptors for it. As I walked by more posters, terms that only hours ago were unfamiliar leapt at me again and again. We won't understand cancer without understanding development, and it was astonishing how, in the year that had passed since the previous meeting, so many new scraps of information had accumulated, the titles laden with that curious terminology: "Fat-Hippo Signaling Regulates the Proliferation and Differentiation of Drosophila Optic Neuroepithelia." (During development Hippo genes help determine the size of organs and have been implicated in certain cancers.) "Fox1 and Fox4 Regulate Muscle-Specific Splicing in Zebrafish and are Required for Cardiac and Skeletal Muscle Functions." (When mutated, they too can propel the growth of malignant tumors.) To draw attention to the findings, a poster would occasionally take a whimsical turn. "1 + 1 = 3" explored the synergistic relationship between two hormones in plant growth. "Where'd my tail go?" was about the Araucana chicken, bred with a mutation that affects its lower vertebrae.

Of all the presentations I saw that day, one lodged deepest in my mind. Weaving through another aisle of posters, titles to my left, titles to my right, I was stopped in my tracks by six little words: "How heart cells embrace their fate." I knew by now that "cell fate" is a technical term and not a philosophical one, that it refers to a fully differentiated cell—one in which the proper pattern of genes has been activated to make a skin cell, a muscle cell, a brain cell. And the subject of this particular study was not the human heart but that of a lowly sea squirt. Still the words rang like poetry.

Just about a mile uptown from the biology meeting was University Hospital, where, not so long ago, Nancy and I had reported for her surgery. Cancer cells are those that rebel against their fate—they hope for so much more—and it made it harder for her knowing that her cancer was in her womb. The ticking biological clock had become a ticking time bomb—anti-life.

The day had begun inauspiciously. The receptionist was brusque, unaware or uncaring that the polite, quiet woman she was talking to was carrying a cancer that could kill her. The admissions clerk was friendly but apologetic. There were no beds available. Like an airline the hospital engaged in deliberate overbooking. Maybe that is inevitable in a major medical complex that also serves as the principal trauma center for the state. In any case Nancy would be entered into the information system as a "floater"—unassigned until, sometime after her surgery, a bed opened up on one of the wards. A floater. The clerk probably didn't know that is police slang for a corpse found facedown in a lake.

The next time I saw Nancy that morning she was lying on a gurney, being prepared for surgery. How bravely she was taking this. As a supervisor stood by, a student nurse stabbed at one of Nancy's veins to draw blood. She missed by a mile and pierced a nerve, causing damage that would persist long after the surgical scars healed. That morning it seemed like a small thing. The anesthesiologist arrived and then the surgeon, offering reassuring words. The double doors opened and my wife was wheeled away.

It was 11:30 a.m., the first Friday in November. We were told it would be a long operation. I found a chair in a quiet corner of the large waiting area, and when I got tired of sitting I would walk the hallways and then find another place to sit. Two hours passed, then three. I didn't want to stray far and miss the surgeon or an assistant emerging with a report. I prayed—if that is what it means to repeat, obsessively, beseeching words in your head. My only god was Einstein's—the laws that govern mass and energy unfolding in the warp of space and time. As my own time slowed, I thought about the strange beauty of science's own creation story. How, long ago on earth, atoms had clasped with atoms to form a multitude of molecules of all different shapes and sizes. How these tiny bits of matter had glommed onto one another in countless configurations, until somewhere along the way one emerged that could duplicate itself. How stray atoms would stick to its nooks and crannies, and what peeled from the mold was another tiny structure identical to the first. And so the process was repeated, matter begetting matter again and again—until somewhere in earth's blue waters the self-perpetuating machinery became caught in a tiny membranous bubble. The ancestral cell was born. It divided and divided, copying itself into child cells that were copied again. All the while molecules inside the cells were subtly changing, mutating spontaneously or from the radioactive background of the earth. But of the new cells that emerged, some were better able to prosper. They could move more quickly toward food or away from danger. Something resembling cancer cells must have appeared in the primordial soup—savage, satanic, proliferating at the expense of the rest. But it would be the cells that could congregate and cooperate that would go on to form multicellular creatures, giving rise to the flora and fauna, the creatures of the earth—these exquisite assemblages in which occasionally a cell, like one inside Nancy, would revert to the wild.

Reverie by reverie afternoon became evening, and still there was no word. I must have covered every linear foot of every hallway on every unlocked floor. I was surprised how easy it was to roam at random without a hospital ID. I walked outside, where orderlies and other staff stood smoking cigarettes. I walked by the emergency room, where victims of knives, automobiles, and guns arrived in ambulances. I walked back up the steps to the surgical floor and sat again. I took out my laptop and tried to work on a book I was writing. It was about Henrietta Leavitt, the woman who in the early 1900s discovered the blinking stars astronomers use as beacons to measure the emptiness of the universe. She died, childless, of stomach cancer. Before long Nancy's brother arrived. The earth had continued to rotate and it was dark outside. The cafeteria closed, and the lights were turned off. We were shooed into a hallway where a family—the only other visitors still on the floor—waited for the outcome of someone else's long surgery.

Finally at 7:30 p.m., eight hours after Nancy had been taken to the operating room, her surgeon emerged, as they do, with his mask hanging loose around his neck. In what is called a modified radical hysterectomy, he had removed her ovaries, fallopian tubes, and uterus, where a tumor—the one that had started this whole thing—had eaten 3 millimeters deep into the endometrial lining and begun spreading into the upper end of her cervix. From there the cancer had maneuvered down one of the round ligaments, which help hold the uterus in place, occupying surrounding tissue as it headed for the right groin—the place where that swollen lymph node had appeared. There it invaded the skin and jumped through the lymph system to nodes in her left groin. Enlarged lymph nodes had also appeared in the pelvic region, two of them angling dangerously close to a vein, but it wasn't yet clear whether these were also cancerous. All of the diseased and suspect tissue had been removed and samples sent for biopsy.

For all of that there was plenty of good news. There was no sign that the cancer had reached any of the organs that sit so close to the uterus: the bladder, the rectum. The cancer had not learned how to form tendrils into the blood system. The operation had gone cleanly and there was no need for a transfusion. Nancy had lost only 300 cubic centimeters of blood, a little over a cup. In his notes for the report that would be typed up a few days later, the surgeon wrote, "Complications: None."

He led us to the recovery room where she lay, just barely awake. She smiled when she saw us and then lapsed back, safe for the night, into unconsciousness. Remembering all this now I am washed over by the sadness my wife had felt about not having children in our lives—a sadness she had tried so many times to explain to me, to get me to feel in my own heart. Now childbearing was no longer an option—with me or with anyone. Instead of an embryo, a cancer was growing inside her, one that like all cancers had borrowed some of the mechanisms of embryogenesis.

## Chapter 7

# Where Cancer Really Comes From

In the 1890s William T. Love, foreseeing an economic boom along the banks of the Niagara River, began excavating a canal. It would skirt past Niagara Falls, allowing boats to travel between Lake Erie and Lake Ontario. More important, the diverted water would be used to generate hydroelectric power. Drawn by a seemingly inexhaustible supply of energy, new industries would spring up. Workers would commute to modern factories from a showcase urban development he would call Model City.

Love's plan depended, in large part, on the need for power-hungry customers to come to the electricity, which in those days was generated in a form pioneered by Thomas Edison called direct current. Direct current could not be carried very far before it faded. The lightbulbs of customers at the end of the electrical lines would be dimmer than those closer to the generating plant. But Niagara's advantage was short-lived. Around the time Love's canal had broken ground, the Serbian inventor Nikola Tesla and his employer, George Westinghouse, introduced alternating current generators and transformers. Before long, electricity, produced at Niagara and elsewhere, could be stepped up to high voltages and transported across the country. That and the great economic panic of 1893 put an end to the Love Canal project, leaving an unfinished ditch about 3,000 feet long and 100 feet wide that residents of Niagara Falls, New York, adopted for swimming and ice skating.

Though Love's project was a failure, other industries, including chemical manufacturers, grew up along the river, and in the years around World War II, the Hooker Electrochemical Company acquired the abandoned canal for use as a dump. Over the next decade, the company disposed of some 22,000 tons of toxic waste, including carcinogens like benzene and dioxin. In 1953 the site, now closed and covered with dirt, was given for a token payment of one dollar to the local school board with the understanding that it was filled with chemical waste. An elementary school was built there anyway and the city envisioned turning part of the old dump site into a park.

During the next two decades land bordering the canal was sold and developed, and in the late 1970s, after a couple of years of unusually high precipitation, residents began to complain of a sickening smell. When an official from the Environmental Protection Agency came to inspect in 1977, he saw rusting barrels of waste that had found their way to the surface. Potholes were oozing waste into several backyards, and it had seeped into the basement of one home. "The odors penetrate your clothing and adhere to your footwear," the official reported. Three days later his sweater still stank. The neighborhood was evacuated, a national emergency declared, and the investigations began.

Whole books have been written attempting to apportion blame among Hooker, the school board, the real estate developers, and the city of Niagara Falls for what everyone agrees was an environmental disaster. (Joyce Carol Oates incorporated the saga into a novel.) Just as difficult has been determining the damages caused by the dump to public health. Early in the crisis, the EPA estimated that people living along Love Canal stood a 1 in 10 chance of getting cancer during their lives just from breathing the polluted air. But several days later the agency admitted to a mathematical error: the increased risk was actually 1 in 100 and far less for people just a few blocks away. Another EPA report found that some of the thirty-six residents who volunteered for tests showed signs of chromosomal damage—more than considered normal. But it was dismissed by a panel of medical experts led by Lewis Thomas, chancellor of Memorial Sloan-Kettering Cancer Center, as "inadequate" and so poorly executed that it "damaged the credibility of science." A later study for the Centers for Disease Control found no excess of chromosomal aberrations.

Cancer can take decades to develop, and those who continued to follow the case awaited the results of a thirty-year retrospective by the New York State Department of Health. With so many variables to juxtapose, studies like this are fraught with uncertainty. Age, sex, and proximity to the canal had to be taken into account. Almost half of the 6,026 residents who were surveyed worked in jobs where occupational exposure might be a risk, and about two-thirds of them had been smokers. About the same proportion drank alcoholic beverages.

When the study was complete, the epidemiologists reported that the birth defect rate for children born to parents who had lived near the canal was double that of Niagara County and also higher than for the rest of the state. Compared with the population at large, slightly more girls had been born than boys—another hint that Love Canal chemicals may have had genetic influences. Despite the hints of teratogenic effects, the study found no convincing evidence that life by the canal had given people cancer. A few types of cancer were slightly more prevalent than expected but the numbers were so small that they were considered within the range of chance. The overall cancer rate was actually a little lower than for the general population.

Birth defects and cancer can both arise from mutations, so why would there be signs of one without the other? It seems plausible that the dividing cells of a developing embryo would be more sensitive to disruptive influences than cells in a fully formed person. And while a single mutation might be enough to derail a developmental pathway, several of these hits would usually be required for a cell in an organ to break away and become cancerous. But even after three decades, the seeming head start provided by Love Canal hadn't been enough to produce an obvious excess of malignancies.

For many of us who grew up during the exuberant beginnings of the environmental movement of the 1970s and 1980s, that outcome was almost beyond belief. We were influenced by *Silent Spring,* Rachel Carson's elegant warning about pesticides and the environment, and scathing polemics like Samuel Epstein's *The Politics of Cancer.* We worried about saccharine and Red Dye No. 2, and later about Alar on apples. We were told of a modern epidemic of cancer— "the plague of the twentieth century"—that was being imposed on the public by irresponsible corporations and their effluents. Food additives, pesticides and herbicides, household cleaners—all of these were said to be corrupting our DNA. We were pawns in "a grim game of chemical roulette," Russell Train, the administrator of the EPA, warned in a story that was picked up by newspapers across the country. "Strange new creatures of our own making are all around us, in our air, our water, our food and in the things we touch. When they hit us, we don't feel a thing. Their ill effects may not show up until decades later, in the form of cancer or even generations later in the form of mutated genes." We were in the midst of what the historian Robert Proctor called "the Great Cancer Wars."

*Ninety percent of cancer is environmental*—we heard that again and again. There was a conspiratorial bent to some of the warnings: the same companies that produced the carcinogenic chemicals also made the drugs used for the chemotherapeutic cures. They were profiting from cancer on both ends. Rhetoric like that was extreme, but the overall message seemed so plausible. Many manufactured chemicals are considered carcinogenic. They can be found among the known and suspected agents listed in the National Toxicology Program's 499-page *Report on Carcinogens.* Depending on the degree of exposure, workers in industries that use or produce these substances take on an increased health risk. As the chemicals diffused through the atmosphere, severe effects on the public were bound to become evident—beginning in the present and escalating year by year with the accumulation of broken genes.

Some of our fears were rooted in a misunderstanding. Epidemiologists define "environment" very broadly to include everything that is not the direct result of heredity—smoking, eating, exercise, the bearing of children, sexual habits, any kind of behavior or cultural practice. Viruses, exposure to sunlight, radon, cosmic rays—these are all defined as environmental. To get a sense of how strongly cancer was influenced by heredity and how strongly by these extrinsic factors, scientists in the 1950s studied populations of black people whose ancestors had been captured by slave traders and moved to the United States and compared them with their relatives who remained in Africa. Liver cancer and Burkitt's lymphoma turned out to be very high among the Africans but not among the American blacks. Lung, pancreatic, breast, prostate, and other cancers were far higher among the black Americans than the Africans. Other researchers found similar patterns. Japanese men were known to have higher rates of stomach cancer but lower rates of colon cancer compared with their counterparts in the United States. When they moved to this country, the situation changed. They tended to adopt the cancers of their hosts and leave the native cancers behind. Since their genes remained the same, factors beyond heredity had to be involved.

By the late 1970s decades of these migrant studies had come to the same conclusion: For 90 percent of cancer cases some kind of external influence was required. Something "environmental." There was a chance of a person getting a head start on cancer by inheriting a damaged gene. But most of the mutations that triggered a malignancy were those acquired during life. That was encouraging news for public health and prevention. But it was often misconstrued to mean that almost all cancer was brought on by pollution, pesticides, and industrial waste. That fit so perfectly with the rest of our worldview that there was little incentive to look deeper. Calmer voices called for a more balanced perspective, but it was the direst warnings that became enshrined in public perception. If we or someone we knew ever got cancer we were quick to wonder whether corporate America was to blame.

There was more to the story than politics and semantics. In 1973, not long after Richard Nixon declared the War on Cancer, the government Surveillance, Epidemiology and End Results Program, called SEER, began collecting data from state cancer registries on incidence and mortality—how frequently people got cancer and how often it killed them. For years the mainstream view had been that except for lung cancer, overall rates were holding steady. But in 1976 when the new SEER data were compared with earlier surveys by the National Cancer Institute, the number of new cases seemed to be escalating abruptly, even when the aging of the population was allowed for. This appeared to be the vindication so many people sought.

Combining two sets of statistics, compiled from different sources according to different rules, is bound to cause trouble. Early on epidemiologists warned that the comparisons were invalid and no conclusions should be drawn—that there was no evidence of a cancer epidemic. To get a clearer idea of what the public was facing, the U.S. Office of Technology Assessment commissioned a study by Richard Doll and Richard Peto, two Oxford University epidemiologists who had made names for themselves by establishing the link between cigarettes and cancer as well as the carcinogenic effects of asbestos. It would have been hard to find two more accomplished scientists in their field.

To begin with, they had to decide which numbers to trust. Although they were improving, statistics on cancer incidence—the number of new cases occurring in a population—were not yet dependable. What appeared to be more new cancers might be the result of better diagnostics, more accurate medical records, and an ever-increasing proportion of the population seeking and receiving medical care. Death certificates from earlier in the century were also suspect. Doctors might acquiesce to a family's request that the stigma of cancer not be entered into the public ledgers. Mistakes in both record keeping and diagnosis were often made. Someone who died of lung cancer might be listed as a victim of pneumonia. A death from an undiagnosed brain tumor might be attributed to senility. A patient might be recorded as dying of cancer when the cause was really something else. The situation improved in 1933 when states began reporting deaths to a central registry, and midway through the century a standardized classification scheme was put in place. (Cervical and uterine cancer had been lumped together, and Hodgkin's lymphoma, a blood cell malignancy, was misconstrued as an infectious disease.) Starting with 1950 and using death rates as the best available approximation for the prevalence of cancer, the authors produced an intricate analysis spanning more than a hundred densely filled pages of words, tables, and graphs and six meticulous appendices. In addition to their own calculations they also reviewed the findings of more than three hundred other studies.

Since it was published in 1981, Doll and Peto's "The Causes of Cancer" has become one of the most influential documents in cancer epidemiology. It concluded that most cancer, by far, is "avoidable"— brought on by factors that, to a great extent, are within the grasp of human control. In 30 percent of cancer deaths, tobacco was a cause. For diet the proportion was 35 percent, and for alcohol it was 3 percent. Some 7 percent of fatalities involved "reproductive and sexual behavior," which included delaying or forgoing the bearing of children and promiscuous sex. (Having multiple partners was recognized as a risk for cervical cancer, although it was not yet known that the agent was human papillomavirus.) Another 10 percent of cancer was tentatively attributed to various infections and 3 percent to "geophysical" phenomena: exposure to the ultraviolet components of sunlight and naturally occurring background radiation from soil and cosmic rays. For deaths by artificially produced carcinogens, including radioisotopes, the percentages came out very low: 4 percent from occupational exposure, 2 percent from air, water, and food pollution, 1 percent from the side effects of medical treatment (including x-rays and radiotherapy), and less than 1 percent from either industrial products like paints, plastics, and solvents or food additives. The remainder was of unknown origin with the suggestion that psychological stress or a compromised immune system might be involved. Except for lung cancer, Doll and Peto concluded, "most of the types of cancer that are common today in the United States must be due mainly to factors that have been present for a long time."

What a hard conclusion this was to swallow. Any specific case of cancer will have multiple causes—environmental (in the broadest sense) along with hereditary dispositions and the elusive influence of bad luck. But for the public at large, chemicals spewed from factories or the polysyllabic additives found in foods were apparently only minor parts of the equation. They were a component—"there is too much ignorance for complacency to be justified," the authors wrote—but far more important was how we lived and the effect that had on a cell's natural tendency to break loose and assert its Darwinian imperative. Most telling of all, Doll and Peto found that cancer had not been increasing rapidly, as one would expect if we were being subjected to an efflorescence of newly invented assaults. When lung cancer and other malignancies closely associated with smoking (oral, laryngeal, esophageal, and others) were removed from consideration, and the aging of the population adjusted for, cancer mortality among people under sixty-five had been steadily decreasing in almost every category since 1953. (That also appeared to be largely true for older Americans, but those figures, relying on earlier medical and census reports, were considered less reliable.) The lower mortality was not because we were getting much better at curing cancer, the authors concluded, but because the number of new cases was not escalating. Once SEER became better established and the quality of data improved, they confirmed that there was no alarming rise in the incidence of cancer.

Doll and Peto were not alone in their findings. Two smaller studies, one in the United States and one for the industrial city of Birmingham, England, had come up with similar percentages—with most cancer attributed to smoking and a mix of other so-called lifestyle factors and with occupational exposure responsible for only a few percent. But "The Causes of Cancer" was the most wide-ranging study that had been undertaken. Its conclusions were, of course, what the leaders of industry wanted to hear, and people committed to fighting the problems of industrial pollution began challenging the report. The lifestyle argument was dismissed as a diversion—blaming the victims instead of the perpetrators. While cigarettes were clearly an important influence, maybe a significant number of smokers wouldn't have gotten lung cancer without additional help from polluted air or synthetic carcinogens—some knotty synergistic effect. Whatever was happening with overall rates, the incidence of some cancers appeared to be rising, especially among the aged and minority groups. Maybe what Doll and Peto laid to better diagnosis were really hints of carcinogenic poisons that were steadily accumulating and would erupt in coming years in a devastating outbreak of cancer. When lung cancer rates began rising earlier in the twentieth century, that was also dismissed as an artifact of better diagnostics. Only with time would the true horror we were inflicting on ourselves become clear.

While epidemiologists kept watch for the appearance of a delayed epidemic, Bruce Ames, the inventor of the Ames test, was also coming to question whether synthetic substances were a significant threat. It was Ames who, back in 1973, had used experiments with bacteria to show that carcinogens, most of them anyway, caused cancer by inducing genetic mutations. (Not all carcinogens are mutagens. Some can work more indirectly. By killing esophagus cells and increasing their rate of replacement, alcohol raises the odds of random copying errors.) As his test became established, Ames worried at first about the hazards of what modern man was releasing into the world. His early research helped lead to bans on carcinogens that were being used as flame retardants in children's pajamas and in hair dyes. He helped persuade California to strengthen its regulation of an agricultural fumigant. He became something of an environmental hero. Then he began testing chemicals that occurred in nature, finding that a surprising number also appeared to damage DNA.

It made good evolutionary sense. Throughout time plants have evolved the ability to synthesize chemicals that ward off predators—bacteria, funguses, insects, rodents, and other animals. Ames described some of these natural pesticides in a paper in *Science* in 1983. The black pepper used to spice our food contains safrole and piperine and causes tumors in mice. Edible mushrooms carry hydrazines that are carcinogenic. Celery, parsnips, figs, and parsley have carcinogenic furocoumarins. In chocolate there is theobromine, and pyrrolizidine alkaloids are found in various herbal teas. Over the years Ames continued to keep count. In 1997, he reported that of sixty-three natural substances found in plants, thirty-five tested as carcinogenic. His most striking example was a cup of coffee—nineteen different carcinogens, including acetaldehyde, benzene, formaldehyde, styrene, toluene, and xylene. Altogether, he estimated, people were imbibing ten thousand times as many natural pesticides as manufactured ones. Those seeking chemical causes of cancer, he said, were looking in the wrong place.

In fact he doubted that nature's poisons were really causing much cancer. Often forgotten is that his paper in *Science* also listed numerous antioxidants and other elements in plants that might conceivably provide some protection. It was possible, Ames proposed, that the good outweighed the bad, that on balance eating fruits and vegetables might reduce the incidence of cancer. But no one really knew.

Ultimately Ames's message was that we were worrying too much about both kinds of chemicals, natural and artificial. Half of everything tested, he wrote, was registering as carcinogenic, but that didn't necessarily mean that the substances were dangerous. Suspected carcinogens are administered to rodents using what is called

the maximum tolerated dose—as much as the animals can take
without debilitating effects. This is many times the exposure that
people might receive in the world. There is a logic to this approach.
Suppose that exposing 10,000 people to some chemical results in a
single instance of cancer. For a population of 10 million that is 1,000
potentially preventable cases. To demonstrate the danger you would
have to give the chemical to tens of thousands of mice—an experiment costing tens of millions of dollars. The alternative is to give
megadoses to many fewer animals and see if a significant portion of
them are affected. The problem, Ames said, was that big concentrations of any foreign substance can throw an animal into physical
turmoil. Sensing the damage to its tissues, the body reacts as though
it has been wounded, unleashing the healing process. That involves
the acceleration of mitosis—rapidly generating new cells to replace
damaged ones. With so much DNA being duplicated, the odds of
random mutations would be higher and so would the possibility of
acquiring one of the deadly combinations. In technical terms, mitogenesis increases mutagenesis.

Toxicologists defended the tests as a reasonably good compromise.
And like Doll and Peto, Ames was condemned by his harsher critics
for giving comfort to polluters and diverting attention from a genuine problem. Perhaps environmental poisons are collecting in the
human bloodstream—barely noticed but still adding incrementally
to the background cancer rate. A recent report by a White House
advisory group suggested that animal tests are actually understating
carcinogenicity—the opposite of what Ames has long contended.
The tests are generally done on adolescent rodents that are sacrificed
when the experiment is over. That would miss the effects of prena-
tal and childhood exposures as well as late-developing tumors. The
alternative would be to administer chemicals to pregnant animals
and follow the health of their babies as they grow into adults and die
of their own accord. Also overlooked would be synergistic interactions. It has been estimated that more than eighty thousand novel
substances have been introduced into the world in modern times.
The number of combinations is endless. Only a small fraction of

new compounds are tested—after they have already been suspected of causing cancer. Taking these factors into consideration, the panel gravely concluded that the number of cancer cases associated with industrial carcinogens "has been grossly underestimated."

While many scientists criticized the report for seriously exaggerating the threat of synthetic chemicals and giving unwarranted credence to a maverick view, few would disagree that toxicology testing needs to be improved. The National Academy of Sciences has described how advances in cellular biology and computer science are opening the way to rapid high-throughput assays allowing many more chemicals and combinations of chemicals to be analyzed. Instead of animals, the tests can be done on cells kept alive in laboratory dishes. The hope is that new carcinogens will be identified quickly and measures taken to reduce their prevalence. If all that should come to pass then cancer rates might be lowered further. That can only be good. But it is hard to make the case that the effect would be very large.

As the years have passed, no epidemic has appeared. Adjusted for the aging of the population, the statistics amassed by SEER show that death rates from cancer did rise gradually by half a percentage point a year from 1975 to 1984—smoking no doubt was a factor—and at a slower pace until 1991, but then they began decreasing modestly and have been doing so ever since. Incidence rates tell a similar story, though the picture is a bit more complex. Like death rates they gradually rose from 1975 until the early 1990s with a burst of newly reported cases from 1989 to 1992, when the rate increased by 2.8 percent a year. The biggest driver for the spike appears to have been more assiduous screening for two of the most common cancers. The number of cases of prostate cancer that were detected shot up by 16.4 percent per year before sharply dropping and breast cancer by 4.0 percent. Then incidence rates, like death rates, began their slow decline.

Every year when the National Cancer Institute publishes the

"Report to the Nation on the Status of Cancer," the story has been the same. Evidence that a large percentage of cases can be attributed to lifestyle has also prevailed. Opinion continues to vary on just which elements are the most important, with specific foods—how much red and processed meat is bad, how many fruits and vegetables are good—giving way to the suspicion that lack of exercise and excess weight are far more to blame. A twenty-five-year retrospective on "The Causes of Cancer" still attributed 30 percent of cancer to tobacco. Obesity and inactivity were believed to account for 20 percent, diet for 10 to 25 percent, alcohol for 4 percent, and viruses for 3 percent. A study by the World Health Organization's International Agency for Research on Cancer found comparable numbers in France. Far down on the lists are occupational exposure and pollutants. Other studies have shown similar proportions in the United Kingdom and other industrialized countries.

Throughout all of this, neighborhood cancer clusters, like the ones I'd read about in Los Alamos and on Long Island and saw fictionalized in *Erin Brockovich,* continue to be reported. But in almost every instance they turn out to be statistical illusions, more examples of the Texas sharpshooter effect. Of those that do not, only a rare few have been associated with an environmental contaminant. Over the decades unusual occurrences of cancer among workers have led to the identification of some carcinogens—the link between mesothelioma and asbestos, for example, and between bladder cancer and aromatic amines (substances also found in cigarette smoke). But even occupational clusters are uncommon.

As the rest of the world develops, the same patterns are appearing as those in the West. Poorer countries tend to be dominated at first by cancers that spread through sexual intercourse and overcrowding—those induced by viruses. There is human papillomavirus and cervical cancer, hepatitis B and C and liver cancer, *Helicobacter pylori* and stomach cancer. With better hygiene and the growing use of Pap smears (and more recently HPV vaccine), cervical cancer may begin to recede. But then new cancers will arise to take its place. As women choose to have fewer children and their better nourished daughters begin menstruating at an earlier age, there may be more estrogen-driven cancers of the uterus and breasts. Education, vaccines, better sanitation—these also push down cancers of the liver and stomach, but at the same time colorectal cancer increases as more people move from the fields to the cities and become slothful. They go from being undernourished to overnourished with all the nutritional imbalances that can come with a modern diet. The cancers of poverty give way to the cancers of affluence. Prostate cancer, a disease of old men, becomes a problem when life expectancy rises into the seventies and eighties. Lung cancer increases as the cigarette companies migrate to less discriminating markets. Industrialization brings with it new dangers of occupational exposure.

Everything doesn't fit into a neat picture. Cancer rates might appear higher in one country than another because of the availability of screening tests. Cancer in urban areas is more likely to be noticed than cancer in the countryside. Beyond the statistical uncertainties, a mix of ingredients—diet, genetics, and cultural practices—can cause surprising variations. The prevalence of mouth cancer in India may come from the chewing of betel nuts and, of all things, reverse smoking—with the lit end of the cigarette inside the mouth. Drinking scalding hot maté may explain the high rates of esophageal cancer in some South American countries. Japan, an affluent society, still leads the world in the rate of stomach cancer. The reason is often laid to diet—a cultural preference for salty fish. Breast cancer in Japan is low for such a developed nation but it is rapidly catching up.

One day, trying to absorb all of this, I holed up in my office and began unpacking the most recent SEER statistics. Concentrating on overall cancer rates can smear over some interesting details, and I wondered what might be lurking underneath. The prime mover in driving down the numbers has been a decline or leveling off in what are by far the most common cancers—cancer of the prostate in men, cancer of the breast in women, and lung and colorectal cancer in both women and men. At the same time, the cancers that appear to be rising—melanoma, for example, and cancer of the pancreas, liver, kidney, and thyroid—are among the rarest. The annual incidence of pancreatic cancer is 12.1 cases per 100,000, compared with 62.6 cases for lung and bronchial. Year by year the figures fluctuate ever so slightly. With numbers so low, it can be difficult to tell if the increases are real or illusory—artifacts created by better reporting and early detection.

That is one of the gnawing difficulties of epidemiology. The scarcer the cancer the more subject the numbers are to random fluctuations, the statistical equivalent of noise. Childhood cancers are among the very rarest, ranging in incidence from 0.6 cases per 100,000 for Hodgkin's lymphoma to 3.2 for brain and nervous system cancer and 5.0 for leukemia. Death rates from these malignancies have fallen to about half of what they were just a few decades ago—one of medicine's great triumphs. But trends in incidence—how many children get cancer in the first place—are almost impossible to decipher. While there is slight evidence of an overall increase, it's very hard to tell. A rise from 11.5 total cases per 100,000 in 1975 to 15.5 in 2009 looks scary. But for the years in between the numbers jump all over the place. The rate was nearly the same, 15.2, back in 1991. The following year it was down to 13.4 and eleven years later, in 2003, it was 13.0. The year after that it was 15.0, then 16.4, then 14.2. What will it be next? You might as well flip a coin.

Every cancer tells a different story. For many years lung cancer declined among men because of the delayed effects of giving up cigarettes. Women started smoking later in the century and so their rates continued to climb. Only recently have they taken a downward turn. A spike in breast cancer in the last quarter of the twentieth century—including the tiny, slow-growing in situ tumors that some doctors don't think should be classified as cancer—may be explained both by better diagnosis and earlier menarche. The recent improvement in the numbers may be partly because of a drop in the use of hormone replacement therapy during menopause. Rising rates of melanoma, which began long before the discovery of the ozone hole, is often attributed to the popularity of sunbathing, tanning salons, and skimpier clothing that protects less flesh from ultraviolet rays. Another reason may be international travel. People from northern climes with lighter skin are now more likely to spend time in sunnier places. What may appear to be a climb in childhood malignancies, the National Cancer Institute suggests, is probably because of better imaging technologies and the reclassification of some benign tumors as malignant. Childhood obesity may conceivably be involved.

You can parse the numbers as finely as you like. Digging into the voluminous SEER figures, one can break out individual cancers by sex, age, race, and geographical locale. Choose a combination of demographics, and different cancers zigzag up and down. Cancer occurs more often in black men than in white men—but less often in black women than in white women. Crack the numbers open further and prostate, lung, colorectal, liver, pancreatic, and cervical cancer are all higher in black Americans, while their rates are lower for skin and uterine cancer and for malignant brain tumors. Darker skin pigments offer protection from sunlight. But the other discrepancies are harder to untangle. Many minorities might be expected to suffer from poorer nutrition, higher rates of smoking and alcoholism, and lower quality medical care—and to live in more polluted areas and work at riskier jobs. But Hispanics, American Indians, Alaska Natives, and Pacific Islanders get significantly less cancer than blacks or whites. There are so many variables involved.

Burrow deeper and more incongruities arise. For all races the incidence of brain cancer ranges in recent years from 4.23 cases per 100,000 in Hawaii to 7.54 in Iowa. That might raise suspicions of an agricultural influence. I wondered what was happening next door to Iowa in Kansas and Nebraska, but these states don't participate in SEER. For liver cancer Hawaii tops out at 10.68, with Utah at the bottom with 3.94. Is that because of teetotaling Mormons or a difference in the prevalence of hepatitis virus? Hours later, wading out of the numerical morass, I despaired of ever making sense of it all. How much easier cancer would be if it were obviously driven by chemical contaminants. Instead there is a muddle of many little influences. High among them is entropy—the natural tendency of the world toward disorder. Of the multiple mutations it takes to start a cancer there is no way to know which was caused by what. Or, in the case of spontaneous mutations—copying errors—if there was a cause at all.

I imagined an army of clones, genetically identical, going through life under the same conditions in the same geographic locales. They would eat the same foods, engage in the same behaviors, and some would die of cancer by the time they were fifty or sixty while others would succumb decades later to something else. As Doll and Peto put it, "Nature and nurture affect the probability that each individual will develop cancer." But it is luck that determines which of us really do.

## Chapter 8

# "Adriamycin and Posole for Christmas Eve"

Among the chemicals on the National Toxicology Program's list of carcinogens is a simple-looking molecule called cisplatin. It is formed when a platinum atom bonds with two chlorine atoms and two ammonia groups. First synthesized in 1844 by an Italian chemist who was experimenting with platinum salts, cisplatin received little attention for more than a century. Then in the early 1960s it was found to have powerful biological effects.

Like so many scientific discoveries this one was serendipitous—a foray into one hypothesis veering unexpectedly in another direction, answering questions no one had known to ask. In his laboratory at Michigan State University, Barnett Rosenberg was exploring how cells behaved in the presence of electricity. He had been struck by how much the stringy, stretched-out shape of a cell undergoing mitosis resembled the field lines that appear when a magnet is held beneath a sheet of paper sprinkled with iron filings. The means by which a cell divides were poorly understood, and he wondered whether some electromagnetic effect might be involved.

Reducing the problem to simpler terms, he placed two metal electrodes in a dish of single-celled organisms, *Escherichia coli,* and applied an electrical current. Before long the bacteria stopped dividing. Each one, however, continued to elongate, producing new protoplasm that extended spaghetti-like until the cell was some three hundred times longer than it was wide. He turned off the current and the cells began dividing normally again. It was like having his finger on a mitotic on-off switch.

Decades later he still remembered the moment: "God, you don't often find things like that," he said. He immediately began thinking about cancer. "If we could control the growth of a cell with an electric field, we could control some cells with a frequency of one sort, other cells with a frequency of another, and then we could attack a tumor by choosing a unique frequency and affecting only the tumor cells and not normal cells." But then came a bigger surprise. It wasn't electricity that was interfering with mitosis. The electrodes that had been used in his experiment were made of platinum, an element he had chosen specifically because it was chemically inert. But through the process of electrolysis some platinum ions were getting into the solution where they combined with other atoms to form cisplatin.

Rosenberg went on to test the molecule's effects on metazoans, creatures like us that consist of many cells. Just a pinch of pure cisplatin was enough to kill a mouse. But in very dilute doses it would cause sarcoma tumors to shrink. Cisplatin also had the power to arrest other cancers, and over the years scientists discovered how that works. Before a cell can reproduce, the double helix must relax its windings so that the molecular information can be copied and passed on to the next generation. Cisplatin caused bridges to form between the two helical strands. This chemical straitjacket blocks mitosis and sends the cell into turmoil. It tries to recover by dispatching DNA-repairing enzymes. When that fails, apoptosis is initiated and the cell destroys itself. Cisplatin can affect any cell in the body, but since cancer cells divide at a faster rate they bear the brunt of the attack. Once the cancer is destroyed, the rest of the body stumbles, as best it can, back to health.

After clinical trials in the 1970s to determine how much cisplatin you could give people without killing them, it was approved by the Food and Drug Administration. It became known as the penicillin of cancer. Because of its effect on other rapidly dividing cells—hair follicles and cells in the gastrointestinal lining and bone marrow—there were sickening side effects. Patients would suffer a bone chilling nausea and their hair would fall out. Kidney and nerve damage might occur, and since cisplatin monkeyed with a cell's DNA it raised the risk of causing a secondary cancer alongside the one the oncologists had been enlisted to treat. The trade-off was usually worth it. For testicular cancer the cure rate approached 100 percent. Other tumors were less responsive, but the chemical, often combined with radiotherapy, could slow cancers of other organs and extend lives. Sometimes it could save them.

Cisplatin, we learned in the days after Nancy's surgery, was one of the agents that would be used in an attempt to kill any remaining metastases that might be hiding inside her, capable of smoldering for many years. She would also get doxorubicin, which like cisplatin operates by interfering with the replication of DNA. Doxorubicin has its own curious tale. The ruby in its name comes from its origin as a red pigment produced by a strain of bacteria. The microbes were discovered inhabiting soil in Italy, and the drug is also called Adriamycin, after the Adriatic Sea. A pretty name, but it too is on the official list of suspected carcinogens. In addition to its nauseating side effects it can push down your white blood cell count, increasing vulnerability to infections. Worst of all it can damage the heart, with reports that the risk increases when Adriamycin is combined with paclitaxel, another mitosis inhibitor that Nancy would be getting. None of this is worse than being dead. Paclitaxel (or Taxol) was originally isolated from the bark of the Pacific yew tree, *Taxus brevifolia*. This discovery was not serendipitous but the outcome of a government program to systematically screen thousands of plants to find substances that were cytotoxic but tolerable—even if just barely—to the human body. That is the brutal nature of chemotherapy. The first chemo agents were derived from mustard gas, whose antimitotic effects were discovered in victims of chemical warfare. Mustargen, which is used against Hodgkin's lymphoma and other cancers, is also called nitrogen mustard and is covered under the 1993 Chemical Weapons Convention.


Every tumor is unique, an ecosystem of competing cells that are constantly evolving, adapting to new threats. Striking a cancer with a combination of different drugs increases the odds of killing it. Nancy's three-pronged onslaught would be especially fierce. The source of her metastasis was initially believed to be endometrioid adenocarcinoma, the most common uterine cancer and one with a pretty good survival rate. But when the postsurgical report came back from the pathology lab, the story became more complicated. Of all the lymph nodes that had been removed only two appeared to be cancerous, and the adenocarcinoma that had been found in her endometrium was judged to be low grade, meaning that the cells had not undergone very many mutations and remained well differentiated. For the most part they still resembled endometrial cells. Invasion into the uterine lining was superficial. None of that made sense. How would so weak-willed a cancer have metastasized so quickly?

The answer seemed to lie in a polyp, a centimeter in size, that had also been excised from the endometrial tissue and biopsied. These cells were much less differentiated and resembled what pathologists call a papillary serous tumor, a type often found in ovarian cancer and one of the most pernicious kinds. But neither the surgeon nor the pathologist saw signs of cancer in the ovaries, which had been removed during the hysterectomy. What had marched with such determination down the round ligament and into the inguinal area was apparently a very rare cancer called uterine papillary serous carcinoma. What little has been published about it can hardly be more discouraging: "UPSC has a propensity for early intra-abdominal and lymphatic spread even at presentation," one oncologist has written. "Unlike the histologically indistinguishable serous ovarian carcinomas, UPSC is a chemoresistant disease from its onset. . . . The survival rate is dismal, even when UPSC is only a minor component . . . and widespread metastasis and death may occur even in those cases in which the tumor is confined to the endometrium or to an endometrial polyp." This new diagnosis, "mixed adenocarcinoma with areas of papillary type intermediate grade," was not clear-cut. The cells in the nodule lacked one of the familiar characteristics—small protuberances or nipples that the pathologist called papillary fronds. But every cancer is different, and UPSC was the pigeonhole with the closest fit.

Looking back years later at the medical records, I see that there were hints of UPSC, or something like it, almost from the start—a sentence in the very first pathology report noting that the cells examined just after the lump appeared had a "micro papillary architecture." If the doctors had suspected from that observation that UPSC was a possibility, they didn't tell us. It was strange that this was the malignancy growing inside her. UPSC is typically a cancer of older, thinner women striking long after menopause and is especially common among African Americans. It is not believed to be tied to increased estrogen exposure and the matter of not bearing children. "There are no risk factors," as two authors bluntly put it. According to one article, as few as 5 to 10 percent of women with stage 4 UPSC—what Nancy had—were alive after five years.

After we read the prognosis, I found an essay, "The Median Isn't the Message," which Stephen Jay Gould, the evolutionary biologist, had written after he was diagnosed at age forty with mesothelioma. This rare cancer, associated with exposure to asbestos, usually affects the tissue surrounding the lungs. Gould's was in his peritoneum, the lining of the abdominal cavity. Once he had recovered from surgery and was beginning chemo, he began researching like mad, quickly discovering that the cancer was considered incurable and that the median mortality after diagnosis was eight months. On its face that would suggest that he would probably die within a year. But Gould began unpacking the statistics. The median, as he explained in his essay, is very different from the mean, or average. It is the halfway point between a range of numbers. If you have a group of seven people and are told that the median height is five foot eight then you know that three of the people are shorter and three are taller. What that doesn't tell you are the extremes. The range of heights might be typical, clustering around the median. But you could also have some abnormally short people all under five feet high, or human beanpoles, or any mix of these and still come out with a median of five foot eight, as long as that was the height of the middle person in the group.

With age of survival, Gould assured himself, there was more likely to be an excess of giants than midgets. The lowest number you could possibly have is zero—the patient is diagnosed at death—but the highest number was essentially open-ended. Plotted on a graph with eight months as the midpoint, the distribution would be asymmetrical: squeezed on the left side between zero and eight but stretched out to the right, including survival times of twelve months, twenty-four months, or many more. As he read about the cancer, Gould found that there were indeed people who had survived for several years. He assured himself that he had every reason to believe he was in that long, right-skewing tail. He was young, otherwise healthy, and as a Harvard professor had access to the best medical care, including a new experimental treatment. "All evolutionary biologists know that variation itself is nature's only irreducible essence," he wrote. "Variation is the hard reality, not a set of imperfect measures for a central tendency. Means and medians are the abstractions." Gould ended up way out on the tip of the tail. He lived for almost twenty more years, dying in 2002, a year before Nancy's diagnosis, of a metastatic lung cancer that his doctors said was unrelated. Nancy was not an abstraction. She was young, healthy, her doctors appeared to be among the best. We hung on to that notion as she began her chemotherapy.


In December, just before Nancy's first session, her father died of the stroke that had brought her home to Long Island just three months before. She wanted with every cell of her body—that is how she put it—to go back there, but the doctors advised against it. We watched a videotape of the funeral instead. Thinking back on those days, a year after her chemo and radiation had been completed, she wrote a short essay, "Adriamycin and Posole for Christmas Eve."

As it opens it is December 22 and she is beginning the second round of what will be seven two-day sessions of intravenous infusions, one every three weeks. Christmas decorations hang in the chemo lounge, and at the nurse's station there is a gingerbread house that had been filled earlier in the season with candy and cookies for the patients. It is almost empty now.

To make the many injections as painless as possible, a chemo port has been installed below her right collarbone—a small artificial blister sitting beneath her skin. It is capped with a silicone rubber membrane through which needles can be poked and is connected internally to a plastic catheter threaded into one of her veins. The device will stay in place for the next few months until the therapy is done.

She sees that one of the "first class seats" is vacant—a comfortable leather recliner with a clear view of the Sangre de Cristo Mountains, where we have hiked together so many times. The sky is grey, with hints of a Christmas snow. As she settles in, I pull up a chair and another round of life's new routine begins. First comes the numbing spray to deaden feeling around the port, and then the premeds, fluids, an antinausea drug—all in preparation for the syringe full of doxorubicin, a.k.a. Adriamycin. It reminds her of red Kool-Aid, and that is the color her urine will turn. While this first cytotoxin is being absorbed by her body, she thinks about Christmas Eve, just two days away, when a few friends will stop by our home for a bowl of posole and tamales, a Santa Fe tradition. The nurse arrives with the cisplatin, and Nancy tries to welcome these chemicals flowing in through that hole in her chest as a gift, a lifeline, no matter how sickening they may be. She tries to envision the shock to all those manically dividing cancer cells when their DNA is suddenly jammed shut—all the delicious apoptotic explosions.

It takes four hours in this room for the day's drugs to be administered. Then we drive home for the night and return the following day for the paclitaxel and four more hours of sitting. Late in the afternoon when the nurse approaches with a shot of Neulasta, which stimulates bone marrow to replace the white blood cells killed by the chemo, we know she has made it through session number two. Three weeks to recover and then back again.

The first nights of these interludes were the hardest. She would awaken in the dark, sometimes so quietly that I didn't hear her rise to go to the bathroom. One morning she told me that she had felt so weak that she lay on the bathroom rug for a while before returning to bed. Why didn't she call out to me, and how had I slept through that? I read years later that because of the toxic effects of chemo drugs, family members are advised to sleep separately and not to share a bathroom. We didn't know that, and I don't think I would have cared.

Early on Christmas Eve she was feeling a little better, and before our guests arrived we left our house to walk down dark, unpaved streets lined with farolitos, the traditional little lanterns made with paper bags, sand, and candles. They light the way for the Christ Child, or so the legend goes. We stopped at one of the bonfires, the luminarias, to warm our hands and legs. Her bones ached from the Neulasta. We avoided Canyon Road, which was already becoming crowded, staying on the side streets. When we reached Acequia Madre, the narrow lane that runs along the town's old irrigation ditch, we came upon something we had never seen. Inside the schoolyard was a man launching flying farolitos—tissue paper balloons fired by candles that ascended and then self-immolated in the sky. I was enough of a traditionalist to feel that this modern touch was an intrusion. I could count on Nancy to see the good in it.

We approach the magician, watch the assembly, and suddenly a light floats up like a miniature paper hot air balloon. Amazing! We follow until the light rises beyond our view. Then, another. There is no way for the approaching one to miss this path of light.

Our guests will arrive in an hour. Suddenly, I can't wait to build our own fire, eat and visit. We bound up the hill toward home. Overhead, I see a bright light in the sky—what's that? It's moving away from me slowly. Can it be? The flying farolito continues its rise, glowing as if it will never go out. I watch, knowing my father can see it too.

Three weeks pass—I don't remember doing anything for New Year's—then back to chemo again. How quickly the unthinkable becomes routine. For all her acceptance of this random blow to life and her gratitude toward the doctors, Nancy questioned everything, and I was there to help with the research. Should she be getting topotecan? She had read that the drug had been used against papillary serous carcinoma and that the response rates of doxorubicin and cisplatin were less than desired. Or did the addition of the paclitaxel tip the scales? "Cisplatin/adriamycin superior without question," the surgeon quickly replied (he and the oncologist had given us their e-mail addresses). He attached abstracts from three papers from the *Journal of Clinical Oncology* and *Gynecologic Oncology* to compare. I thought of the surgical report he had written—so clear, precise, and literate. These were doctors who kept up with the research and expressed themselves cogently and persuasively.

One day Nancy's oncologist gave us a paper, published just a few months earlier, called "HER2/neu Overexpression: Has the Achilles' Heel of Uterine Serous Papillary Carcinoma Been Exposed?" HER2/neu was a gene that codes for receptors that respond to human epidermal growth factors—signaling molecules that encourage mitosis. It is usually just called HER2. Some breast cancer cells have too many copies of the gene. Instead of two, one from each parent, there are fifty or a hundred and the cell's membrane becomes glutted with receptors. Tens of thousands of receptors is normal. A HER2-positive breast cancer cell might have 2 million. Wildly overacting to growth-stimulating signals, the cells multiply with mad determination. A drug called Herceptin was designed to seek out the receptors and shut them down—in breast cancer and possibly in other cancers too.

Though more precise than the blunderbuss of chemo, these new "targeted therapies" weren't always as targeted as they sound. There was still unwanted damage to healthy cells, and as with other drugs the cancer would evolve antidotes and counterstrategies, mutations that confer resistance. But given what we had been hearing, the possibility described in the new study seemed like unusually promising news. Many UPSC cells, the author found, also overexpressed HER2—even more so than in breast cancer—and they withered in the presence of Herceptin. He wasn't reporting successful results from the clinic—these were in vitro experiments—but another avenue had opened. And almost as quickly, it closed. A diagnostic test was ordered on Nancy's cancer cells but it came back negative. The amount of HER2 was normal. Herceptin wasn't an option, but we wondered what other possibilities might be out there, findings too new to have made it into the journals.

We were helped in the search by some of my colleagues who wrote about science and health for *The New York Times*—Sandra Blakeslee, Denise Grady, Jane Brody, and Lawrence Altman, a doctor who decided early on to write about medicine rather than practice it. Altman was the reporter on call when the *Times* needed a story explaining whatever ailment had befallen the president of the United States. (His name had been mentioned in an episode of *The West Wing* when the fictional President Bartlet held a press conference about his multiple sclerosis.) When we decided to seek an outside opinion from the MD Anderson Cancer Center in Houston, Altman sent an e-mail to John Mendelsohn, the president, and we had an appointment for the last week in January. We were luckier than most in so many ways.

Anyone with cancer can hardly resist the Anderson allure. "Making Cancer History" was its slogan, and an impressive packet of information quickly arrived. A large brochure with photographs of smiling doctors and patients described how far Anderson would go beyond what might be expected at the local hospital. Through the office of Patient Guest Relations and the Patient Travel Services one could book discounted airline tickets with no penalties for last-minute changes. There was a concierge. Hotel rooms were available right on the grounds at the Jesse H. Jones Rotary House International. Maps and parking passes were included in the envelope with instructions for negotiating the vast Anderson campus. "Do not be overwhelmed by our size," patients were advised. "We are here to guide your journey through our hallways."

There was a Learning Center with medical reference books and videos, a Leisure Library if you preferred a good novel. A Craft Room, a Music/Game Room—and all of that was beside the point. People came to Anderson because it operates one of the largest and most respected research centers in the world. If there were new things to learn about UPSC or trials of experimental treatments, Anderson would surely know.

The evening of our arrival we ate a bland but probably healthy supper at the Rotary House restaurant and then returned to our room to wait for the morning. There was a closed-circuit Anderson channel on the television, and when we tuned in it was airing meditation and visualization exercises: Close your eyes and imagine the golden light of health flowing through you. It didn't sound very scientific, but anything that relieved stress could only be good. We were early the next morning for our appointment with the professor of gynecologic oncology, one of the grand old men of the field, who also served as special assistant to the president of the cancer center. By now Nancy had shed all of her thick brown hair, but she looked as pretty as ever in her scarf. Another patient, new to cancer, came over to ask what it would be like when her own hair fell out. Would it happen all at once or gradually? She would soon be worrying about other things.

The medical records and microscope slides had been sent in advance from New Mexico, and the doctor had familiarized himself with the surgical and pathological reports and the chemo protocol. "UPSC—that's a tough one," he said. Nancy went off for a quick medical exam, and when she and the doctor returned we all took seats in his office. He agreed with everything the oncologists were doing in Santa Fe. It was just what he would have done at Anderson. "You're getting state-of-the-art care," he said. We left the building feeling both relieved and a little disappointed. It was reassuring to have his imprimatur. But we had hoped to be bestowed with some new laboratory finding, a promising clinical trial, some kind of Anderson magic.

With the rest of the day to fill, we took a tour of the Lyndon B. Johnson Space Center south of central Houston and saw the old Mission Control Center, the nexus of operations for Apollo 11 when a human first walked on the moon. Anything seemed possible then. Back in the city we visited the Rothko Chapel. Years before when we had lived in New York, Mark Rothko and Jackson Pollock were two of our favorite artists at the Museum of Modern Art. Pollock's drip paintings always left me feeling that I was peering inside the frenetic workings of a human brain—ideas looping and sparking in motions that teetered between order and chaos. Pollock stimulated while Rothko, with his big blurred blocks of color, soothed. Inside the octagonally shaped chapel he had taken this serenity to an extreme: eight walls of enormous black canvases. We stared at them trying to find patterns, some subtle meaning.

## Chapter 9

# Deeper into the Cancer Cell

Things are rarely as simple as they seem, and what appears to be complex may be no more than ripples on the surface of a fathomless ocean. The mechanics of malignancy I was slowly becoming comfortable with—with a single cell acquiring mutation upon mutation until it spirals down the rabbit hole of cancer—was neatly described by two scientists, Douglas Hanahan and Robert Weinberg, in a sweeping synthesis published in 2000 called "The Hallmarks of Cancer." Both authors are respected researchers. Weinberg, a pioneer in the discovery of the first oncogenes and tumor suppressors, would be on anyone's list of the most prominent and original thinkers in his field.

The idea of cancer occurring as an accumulation of mutations to a normal cell goes back decades. But it was Hanahan and Weinberg who assimilated a growing mass of laboratory results and theoretical insights into six characteristics that a cancer cell must acquire as it develops, in its pell-mell version of Darwinian evolution, into the would-be creature called a tumor. It must acquire the ability to stimulate its own growth and to ignore signals admonishing it to slow down. That is where the oncogenes and tumor suppressors come in.

It must learn to circumvent the safeguard of programmed cell death and to defeat the internal counters—the telomeres—that normally limit the number of times a cell is allowed to divide. It must learn to initiate angiogenesis—the sprouting of its own blood vessels—and finally to eat into surrounding tissue and to metastasize.

More than a decade after it was published, "Hallmarks" was still the most frequently cited paper in the history of the prestigious journal *Cell,* which is as good as saying that it may be the single most influential paper on the biology of cancer. Known as the monoclonal theory (a dividing cell and its branching tree of descendants is called a clone), the picture spelled out in "Hallmarks" remains the dominant paradigm, like the big bang theory is in cosmology. Creation began as a singularity—a primordial dot of mass-energy—and ballooned to form the universe. A cancer begins with one renegade cell—it was Weinberg who popularized that term—expanding to form a tumor. With this rough map in place, the two scientists looked forward to a renaissance in the understanding of cancer:

> For decades now, we have been able to predict with precision the behavior of an electronic integrated circuit in terms of its constituent parts—its interconnecting components, each responsible for acquiring, processing, and emitting signals according to a precisely defined set of rules. Two decades from now, having fully charted the wiring diagrams of every cellular signaling pathway, it will be possible to lay out the complete "integrated circuit of the cell." . . .
>
> With holistic clarity of mechanism, cancer prognosis and treatment will become a rational science, unrecognizable by current practitioners. . . . We envision anticancer drugs targeted to each of the hallmark capabilities of cancer. . . . *One day, we imagine that cancer biology and treatment—at present, a patchwork quilt of cell biology, genetics, histopathology, biochemistry, immunology, and pharmacology—will become a science*

*with a conceptual structure and logical coherence that rivals that of chemistry or physics.*

A physics of cancer! In the decade and more that has passed since this immodest prediction, scientists have continued to uncover whole new layers of complications. Inside the biological microchip called a cell there are components inside components and wiring so dense and so fluid that it sometimes seems impossible to tease the strands apart. Moving up a level, what is happening inside a cancer cell cannot be fully understood without considering its place within an intricate communications network of other cells. By the time the "Hallmarks" paper was published, scientists were already finding that tumors are not homogeneous masses of malignant cells—that they also contain healthy cells that help produce the proteins a tumor needs to expand and attack tissue and to plug into the blood supply. This aberrant ecosystem has come to be called the cancer microenvironment, and entire conferences and journals are devoted to understanding it.

Complicating matters further has been the gradual realization that the genetic changes that can lead to cancer don't necessarily have to occur through mutations—deletions, additions, or rearrangements of the nucleotide letters in a cell's DNA. The message can be altered in more subtle ways. Think of what happens during normal development. Every cell in the fetus carries the DNA inherited from its parents—the genetic instructions a body requires to manufacture its many parts. As cells divide and differentiate the entire script remains intact, but only certain genes are activated to produce the proteins that give a skin cell or a kidney cell its unique identity. That much is familiar biology. What hadn't occurred to me is that as the cell proliferates, this configuration must be locked in place and passed on to its progeny.

Scientists have been piecing together a rough picture of how this works. Molecular tags can bind to a gene in a way that causes it to be permanently disabled—incapable of expressing its genetic message.

(The tags are methyl groups, so this process is called methylation.) Genes can also be enhanced or suppressed by twisting the shape of the genome. In the iconic image, DNA's interwoven coils float elegant as jellyfish in lonely isolation. But in the messiness of the cell, the two helical strands are wrapped around clusters of proteins called histones. Methyl groups and other molecules can bind to the helix itself or to its protein core and cause the whole assembly to flex. As that happens some genes are exposed and others are obscured. Alterations like these, which change a cell's function while leaving its DNA otherwise unscathed, are called epigenetic. "Epi-," coming from ancient Greek, can mean "over," "above," "upon." Just as a cell has a genome, it also has an epigenome—a layer of software overlying the hardware of the DNA. Like the genome itself the epigenome is preserved and passed on to daughter cells.

What all this suggests is that cancer may not be only a matter of broken genes. Disturbances to a cell—carcinogens, diet, or even stress—might rearrange the epigenetic tags without directly mutating any DNA. Suppose that a methyl group normally keeps an oncogene—one that stimulates cellular division—from being expressed. Remove the tag and the cell might start dividing like crazy. On the other hand, the production of too many tags might inactivate a tumor suppressor gene that would normally hold mitosis in check. Freed to proliferate, the cell would be vulnerable to more copying errors. So epigenetic changes would lead to genetic changes—and these genetic changes could conceivably affect methylation, triggering more epigenetic changes . . . and round and round it goes.

Outside the laboratory enthusiasm for this scenario is driven both by hope and by fear. Epigenetics might provide a way for a substance to act as a carcinogen even though it has been shown incapable of breaking DNA. But unlike genetic damage, these changes might be reversible. How big a role epigenetics plays remains uncertain. Like everything that happens in a cell, methylation and the modification of histones are controlled by genes—and these have been found to be mutated in different cancers. Maybe it all comes down to mutations after all. On the other hand, a few scientists have proposed that cancer actually begins with epigenetic disruptions, setting the stage for more wrenching transformations.

Even more unsettling is a contentious idea called the cancer stem cell theory. In a developing embryo, stem cells are those with the ability to renew themselves indefinitely—they are essentially immortal—dividing and dividing while remaining in an undifferentiated state. They are agents of pure potentiality. When a certain type of tissue is needed, genes are activated in a specific pattern and the stem cells give rise to specialized cells with fixed identities. Once the embryo has grown into a creature, adult stem cells play a similar role, standing ready to differentiate and replace cells that have been damaged or reached the end of their life. Since healthy tissues arise from a small set of these powerful forebears, why couldn't the same be true for some tumors?

This would be an unexpected twist on the conventional view in which any cancer cell that has acquired the right combination of mutations is capable of generating a new tumor. Imagine if instead the growth and spread of a cancer is driven by a fraction of special cells, those that have somehow become endowed with an intrinsic quality called "stemness." Just as normal stem cells generate skin, bone, and other tissues, the cancer stem cells would generate the variety of cells that form the rest of a tumor. But only the cancer stem cells would have the ability to replicate endlessly, metastasize, and seed another malignancy. How much easier that might make things for oncologists. Maybe chemotherapies fail because they spare the cancer stem cells. Remove these linchpins and the malignancy would collapse.

It is a promising possibility, but the further I ventured into the subject, the more confusing it seemed. Do the other cells in the tumor perform functions like angiogenesis that would aid in sustaining the malignancy? Or are they just filler material? And where would the cancer stem cells come from? Do they begin as normal stem cells (like those that generate skin) that become damaged by mutations? Or are they fetal stem cells that survived into adulthood and then went berserk? Or, like the other cells jostling for position inside a tumor, did they also arise through random variation and selection? Maybe the all-powerful cells began as "ordinary" tumor cells that shed their identity and reverted to this primal form. Some experiments suggest that in the turmoil of a tumor, cells are constantly shifting their identity between cells with stemlike properties and cells without.

As I struggled to fit this all into the big picture, I was relieved to find researchers who seemed as baffled as I was. Some scientists were convinced the hypothesis was the wave of the future, others that it was of limited importance—a footnote to the standard theory. However it all pans out, the underlying view of cancer as a Darwinian process—arising like life itself through random variation and selection—would remain largely unshaken. But as an outsider trying to understand the essence of cancer, I felt daunted by the possibility of even more convolutions.

The place to take in the full sweep of what is happening on the frontiers is the annual meeting of the American Association for Cancer Research, the largest and most important of its kind in the world. It was being held early one spring in Orlando, Florida, and as I changed planes in Atlanta I could already see the ripple effect. Young scientists rushed through the airport carrying long cardboard tubes protecting their posters. Each, unfurled, would describe a tiny piece in the expanding puzzle. Altogether more than 16,000 scientists and other specialists from sixty-seven countries were converging on Orlando, where more than six thousand new papers would be presented—in poster sessions and symposiums—over a span of five days. There were few distractions. Orlando's mammoth convention center and its environs form an insular world of hotels, chain restaurants, and meeting halls, a kind of boring version of Las Vegas.

Inside this air-conditioned bubble I hoped to absorb as much as I could.

While there had been three simultaneous sessions at the modest developmental biology meeting I had attended in Albuquerque, here there were more than a dozen—beginning at 7:00 a.m. and running into the evening with major lectures and educational sessions overlapping and in between. Carrying a copy of the proceedings as thick as a telephone book (or the weightless equivalent on their cell phones), the informavores plotted their hunting strategies. As the clock began to run out on a speaker's 10:30 a.m. talk, there would come a rustling of chairs with people quietly hurrying to a presentation scheduled in another room for 10:45. Geography was a consideration. Going from "Guts, Germs and Genes" (recent findings on the role bacteria play in the onset of some tumors) to catch the end of "Ubiquitin Signaling Networks in Cancer" required a brisk ten-minute walk indoors. Beckoning one floor below was the exhibit area, where pharmaceutical companies with huge steampunk espresso machines tempted passersby—a cappuccino and biscotti in return for listening to a presentation by Merck or Lilly on a new cancer drug. At the Amgen booth, visitors wearing 3-D glasses watched an amazing video flythrough of a tumor undergoing angiogenesis. For more than a decade Amgen had been working on an angiogenesis inhibitor. Combined with paclitaxel in a clinical trial, it extended the lives of women with recurrent ovarian cancer from 20.9 months to 22.5 months, or about forty-eight days.

As I watched the video, I thought of the excitement thirteen years earlier when a Harvard scientist, Judah Folkman, had discovered what briefly appeared to be the makings of a silver bullet. For every mechanism in a cell there is a countermechanism to keep it in check. Angiogenesis is a normal means through which blood is supplied to newly created tissues. Molecules called angiostatin and endostatin, which are naturally produced to inhibit angiogenesis—you don't want new blood vessels growing just anywhere—had shown striking effects in choking off tumors in mice. James Watson, the celebrated molecular biologist, was quoted on the front page of *The New York Times:* "Judah is going to cure cancer in two years." He followed up with a letter to the editor insisting that he had spoken more cautiously to the reporter—and then went on to declare, just as enthusiastically, that what was happening in Folkman's laboratory was "the most exciting cancer research of my lifetime, and it gives us hope that a world without cancer may yet be attainable." Watson was not alone. The director of the National Cancer Institute called Folkman's results "remarkable and wonderful" and "the single most exciting thing on the horizon," before adding the usual caveat that what worked for mice wouldn't necessarily work for people.

It didn't, of course. The experiments were difficult to replicate and later research suggested that some angiogenesis inhibitors might make matters worse—with the tumor fighting back by metastasizing more vigorously toward safer ground. There are now inhibitors on the market, but the results are nothing like what had been envisioned. Used along with the standard blunt-edged poisons, Avastin can add a few months to a patient's life at a cost of tens of thousands of dollars. Side effects include gastrointestinal perforation and severe internal bleeding. Inhibiting angiogenesis can interfere with the healing of surgical incisions and other wounds. Several months after the Orlando meeting, the Food and Drug Administration, weighing the risks and the benefits, revoked approval for Avastin as a treatment for metastatic breast cancer.

Such grim realities seemed far away at the grand opening session, where Arthur D. Levinson, a pioneer in the design of targeted therapies, was honored for "leadership and extraordinary achievements in cancer research." He was cited specifically for his role in developing "blockbuster drugs" like Avastin. Levinson is the chairman of Genentech, which also makes Herceptin to treat the 15 to 20 percent of breast cancers that are HER2 positive—those with an overabundance of the growth-stimulating receptors. For metastatic breast cancer, Herceptin can add a few months to a woman's life. Used in the early stages of the illness, the drug's effects are more striking.

When standard chemotherapy was accompanied by Herceptin, 85 percent of women were found free of the cancer after four years. That compared with 67 percent who had not taken the drug. The trial was stopped early so that women in the control group could benefit (and so Genentech could reduce the time to market). As word spread of the new therapy, breast cancer patients who once dreaded learning that their tumor was HER2 positive—a particularly vicious and aggressive kind—came almost to welcome the news.

No cancer drug, however, is as good as it sounds. Herceptin can also affect healthy cells with a normal number of HER2 receptors, and there is a serious risk of congestive heart failure. Even Gleevec, the "crowning achievement" of targeted therapy, has its dark side. With the drug, chronic myeloid leukemia can almost always be held in check, but Gleevec must be taken indefinitely to keep the cancer from coming back. There are also problems with another class of pharmaceuticals that aim to suppress tumors by strengthening the body's immunological defenses. Immune system boosters called cytokines are infused into the bloodstream—or the patient's own immune cells are removed, modified to enhance their killing powers, and then reinjected. The danger with these experimental therapies is keeping the immune system from becoming so vigilant that it wildly overreacts, mistaking the body itself for an interloper and initiating a catastrophic autoimmune response.

As I pondered what counts as a blockbuster drug, the auditorium was aroused by a fanfare of strings. This was a first for me—a scientific meeting with its own musical theme. Harold Varmus, the director of the National Cancer Institute, was taking the stage. To accommodate an audience of thousands of people, each speaker's image was projected on six sets of double screens—one half for the video of the lectern and the other for PowerPoint slides. The images loomed so large that the speaker himself, off in the distance, appeared comically small, the man behind the curtain in *The Wizard of Oz*. Varmus began with the good news: Overall incidence and mortality rates were continuing to inch a little lower every year.

That, of course, is after adjusting for the aging of the population. The frightening reality, he reminded everyone, is that wave after wave of baby boomers are entering their sixties and seventies—prime cancer time. Even with a modest decline in the amount of cancer per capita, the sheer number of cases will soar. At the same time government research funding was not even keeping up with inflation. "We're not just poor but living in a land of uncertainty," Varmus lamented.

Watching these lavish presentations with their state-of-the-art audiovisual enhancements, I found it hard to think of cancer as medicine's neglected stepchild. All medical research has been threatened by budget cuts. But when you add to the government grants the money that is going toward pharmaceutical research (the justification given for those five-figure drug price tags) and the private dollars raised in telethons and donated by the wealthy hoping to stave off their own death or to memorialize a loved one with a new medical center wing, great resources were going toward understanding cancer in the minutest detail. Would billions of additional dollars soon lead to the drugs, always just beyond the horizon, that would zero in on advanced-stage cancer without the collateral damage of chemo and radiation, buying not just weeks or months but an actual cure? Would death rates fall as precipitously as they have for heart disease? Would people stop lamenting that we are losing the War on Cancer?

There is so much money to be made in the fight, and I was taken aback by how many top university researchers had a hand in the commercial world. Elizabeth Blackburn, who was stepping down as president of AACR, had won a Nobel Prize for her research on telomeres and telomerase. She was also a founder and chairman of the advisory board of an enterprise called Telome Health, Inc. Throughout the week every presentation began with an obligatory slide disclosing any conflicts of interest. There was clearly some resentment over the requirement. Some speakers flashed the words so quickly that they were impossible to read. I was reminded of those television car commercials where a comically sped up voice rapidly spews out the fine print and disclaimers. One plenary speaker hurriedly said that she had lost her slide. (It would have indicated that she and her husband were cofounders of a publicly traded pharmaceutical company that is developing targeted cancer therapies.) Other speakers proudly declared, often to applause, that they had nothing to disclose, and one said that his biggest conflict of interest was that for twenty-five years he had worked on a skin cancer treatment "and therefore I really want this stuff to work."

Varmus is one of the giants of medical science, sharing his own Nobel Prize with J. Michael Bishop for their pioneering work on viruses and oncogenes. He seemed glad to get money matters out of the way so he could move on to the science and some of the most perplexing questions it faced: Why is it that some cancers—testicular, for example, and some leukemias and lymphomas—can be killed by chemo alone while others are stubbornly resistant? What are the biological mechanisms that account for obese people having a higher cancer risk? Why do patients with neurodegenerative diseases like Parkinson's, Huntington's, Alzheimer's, and fragile X appear to be at lower risk for most cancers? Why do the body's tissues differ so dramatically in their tendency to develop cancer? As I listened, it occurred to me that I had never heard of cancer of the heart. (It does occur but is extremely rare.)

For the rest of the morning other luminaries stepped forth to speak about the future, each preceded by the rousing melodic fanfare and the disclaimer slide. With the latest technology researchers are sequencing the genomes of cancer cells, far more rapidly than had seemed possible even a few years ago. By comparing the tumor genomes with those of normal cells, they are seeing on a finer grain than ever the mutations that can produce a malignancy. Some of the results have been surprising. According to the common wisdom it typically takes half a dozen or so damaged genes to tip a cell. But two cases of the same kind of cancer (breast cancer, say, or colon cancer) needn't arise through the same combination of genetic alterations. Genomics research suggests that for some cancers dozens and even hundreds of mutations may potentially be involved. Of the approximately 25,000 genes in the human genome, at least 350 have been identified as possible cancer genes—ones that can be altered in a way that confers a competitive advantage. According to some predictions, the number may eventually run into the thousands.

"Cancer is not a disease. It's a hundred different diseases"—how many times has that been said? Now the talk is of cancer as tens of thousands of diseases each with its own molecular signature. One day, as these technologies develop, scientists may be able to routinely analyze the unique characteristics of every individual cancer and provide each patient with a personally crafted therapy. It is a lot to hope for.

We left the auditorium, the thousands of us, and diffused throughout the cavernous spaces of the convention center. Every lecture room and every corridor of posters offered more elaborations on the cancer theme. There was the phenomenon of polarization—the way a healthy cell can tell its front from its back. This allows epithelial cells to orient themselves within a tissue so that hair, scales, and feathers all lean the same way. During mitosis a cell must polarize, portioning out its contents before it splits into two identical cells. A migrating cell is exhibiting polarization when it transports its proteins in a way that keeps it moving forward and not backward, as though riding on its own conveyor belt. Some of the molecular circuits involved in polarization have been uncovered, and in a cancer cell they are among the things that can go askew. Whether that is a symptom or a cause of the malignancy is another of the unknowns.

While that question was being pondered, researchers in another room were discussing the many different kinds of cell death. Switching off apoptosis is an established hallmark of cancer, and chemotherapy typically works by forcing apoptosis back on. But there are also autophagy (the cell eats its own insides), entosis (a cell cannibalizes its neighbor), and necroptosis, which like apoptosis involves molecules called death receptors and RIPs (the epitaph stands for "receptor-interacting protein"). Maybe these too can be manipulated in controlling cancer. There is a *Journal of Cell Death,* and a woman in the audience was wearing a black T-shirt with the cryptic words "Cell Death 2009: The Unplugged Tour." So many little subcultures even in the cancer world.

Other speakers pondered the mystery of why cancer cells change their metabolism from aerobic to anaerobic, voraciously consuming glucose in a phenomenon called the Warburg effect. This less efficient way to use energy would help them survive in the oxygen-starved reaches deep inside a tumor. But the cells also make this transformation when there is plenty of oxygen available. One reason might be that the altered metabolism allows them to take in more of the raw material they need to build new parts and proliferate. There were lectures on the ways in which a cancer cell can elude destruction by the immune system—or turn it to its own uses, attracting macrophages as allies in the cause. The slow burn of chronic inflammation is somehow involved with many diseases—rheumatoid arthritis, Crohn's disease, Alzheimer's, obesity, diabetes—and it also plays a role in cancer. Stomachs inflamed by an immune response to *Helicobacter pylori* bacteria or livers inflamed by hepatitis virus are more likely to become cancerous. But how much is cause and how much is effect? The chemical circuitry is still being uncovered. A full session was devoted to the question of how molecules called sirtuins, which have been implicated in the aging process, also play a role in inflammation, obesity, and therefore in cancer.

In the end what all of biology comes down to is genes talking to genes—within the cell or from cell to cell—in a constant molecular chatter. I hadn't considered, however, that the genes in human tissues can also talk to the genes residing in the microbes that occupy our bodies. Maybe that should have been obvious. Our skin and our digestive and respiratory tracts are teeming with bacteria. Many of them play a symbiotic role—bacteria in the gut secrete enzymes that aid in digestion. The genes inside these single-celled creatures transmit signals from microbe to microbe, and they can also exchange signals with human cells. Although we think of the bacteria as passengers, they outnumber our own cells by about ten to one. Even more impressive, the total number of microbial genes each of us harbors—the microbiome—outnumbers our human genes by 100 to 1. There is even a Human Microbiome Project to sequence the genomes of these cellular free agents. Cancer is a disease of information, of mixed-up cellular signaling. Now there is another realm to explore.

The genome, the epigenome, the microbiome—scientists also now speak of the proteome (the entire set of proteins that can be expressed in a cell) and the transcriptome (all of the RNA molecules of various sorts). There is the metabolome, lipidome, regulome, allelome . . . the degradome, enzymome, inflammasome, interactome, operome, pseudogenome. . . . The exposome is everything in the environment we are exposed to and the behaviorome includes the lifestyle factors that may alter our risk of cancer. The bibliome is the endlessly expanding library of papers on everything scientific, and the curse of this age of microspecialization and the proliferation of "'omics" is to separate the ridiculome from the relevantome.

As I scribbled in my notebook or walked the hallways mulling some strange new idea, I thought of how much has changed over the years in our understanding of cellular biology. I remembered the thrill of reading James Watson's *The Double Helix* during a backpacking trip in college and, later on, sitting by the fire in a mountain cabin, devouring the three-part *New Yorker* series excerpted from Horace Freeland Judson's magnificent book *The Eighth Day of Creation: Makers of the Revolution in Biology.* Molecular genetics seemed as clean and crisp as structures assembled from Lego bricks. For all their power to create and govern life, genes were made from combinations of just four nucleic acid letters: G, C, A, and T. Each had a unique contour, and these patterns of bumps and grooves were copied from DNA to messenger RNA and then ferried to the ribosomes, the cellular structures that used the information to make proteins.

At these foundries other molecules called transfer RNAs acted like adaptor plugs matching each triplet of nucleic acid letters to a particular amino acid—the twenty different units that, arranged in a certain order, became a particular kind of protein. These proteins include the enzymes that help make the genetic machinery run. The crowning simplification of the theory was what Francis Crick called the "central dogma": DNA to RNA to protein.

The complications were soon to follow. Not every bit of DNA was part of the protein code. Some sequences were used for making the messenger RNA and transfer RNA. Others served as control knobs, turning the volume of a gene up and down to modulate the production of its protein. With all of this intricate, interlocking machinery, you could almost entertain the fantasy that the whole thing was the product of an engineer. But nature was so much messier. Genes, for example, were not continuous. They were interrupted by scraps of gibberish. As the genetic message was reprinted into the messenger RNA, these blemishes (the introns) had to be edited out. They were accidents of evolution and of entropy. In fact, of the entire genome only a small percentage appeared to serve a purpose. The rest came to be known as junk DNA—a hodgepodge of detritus, genes that had become crippled and discarded over the course of millions of years. Some of these pseudogenes had been smuggled in by viruses. Others were created when a real gene was mistakenly copied and pasted elsewhere in the genome. With no compelling reason to get rid of the debris, it was carried along, generation by generation, for the ride.

It seemed barely conceivable that so much of the genome sat silent and inert. In its incessant tinkering, evolution would surely find new purposes for the discarded parts. Early in the 1990s, scientists began to notice a new kind of RNA produced by the junk DNA. When they latched onto a messenger RNA, these molecules kept it from delivering its information. Because of their small size they were named microRNAs (in the lexicography of cellular biology, terms like this are squished together). They came in different varieties, and as they increased or decreased in number they regulated the production of various proteins. Like almost everything else in the cell they were bound to play a role in cancer. Suppose there was a microRNA whose role was to block the expression of a growth-promoting oncogene. If the cell produced too little of this regulator, that would encourage proliferation. An excess of another kind of microRNA might result in the stifling of a tumor suppressor. In fact just one of these molecules might regulate several different genes, leading to tangles of entwined effects. Mutations to the junk DNA had been thought to be harmless. But if they upset the balance of microRNAs they could nudge a cell closer to malignancy.

The closer scientists looked, the more varieties of RNA were found. Some of these molecules may be flotsam and jetsam—broken pieces left over from the day-to-day running of the cellular machine. But others seem to be there for a purpose. LincRNA (for large inter-vening noncoding), siRNA (for small interfering), and piRNA. That means Piwi-interacting, and Piwi (for P-element induced wimpy tes-tis) is another of those genes with silly names. There is Xist RNA and Hotair RNA. Wherever their names come from, the important idea is that these molecules too can play a role in regulating cellular chemistry. They might cause runaway cell growth if their balance is upset. A few scientists, reluctant to jump on the bandwagon, think the importance of the new RNAs has been overblown. Others think they herald a revolution. Declaring that "the central dogma is bro-ken," a Harvard scientist speaking in Orlando described a sweeping new theory in which genes talk to pseudogenes in a new language whose letters consist of these exotic RNAs. If he is right then there is another code to be deciphered. Only then can we truly understand the cellular circuits and how they can go wrong.

Junk that is not junk. Genes—99 percent of them—that reside in our microbes rather than in our own cells. Background seemed to be trading places with foreground, and I was reminded of what happened in cosmology when most of the universe turned out to be made of dark matter and dark energy. Yet for all the new elaborations, the big bang theory itself was left standing. It wasn't so clean and simple as before, but it provided the broad strokes of the picture, a framework in which everything, aberrations and all, made sense. The same appeared to be happening with the hallmarks. One presentation after another in Orlando included a much-copied PowerPoint slide illustrating Hanahan and Weinberg's six canons. Without that touchstone all would have been chaos. Just the month before, the two scientists had published a follow-up: "Hallmarks of Cancer: The Next Generation." Looking back on the decade that had elapsed since their paper, they concluded that the paradigm was stronger than ever. Certainly there were complications. In the microchip of the cancer cell what had appeared to be a single transistor might turn out to be a microchip within the microchip hiding more dense circuitry of its own. Stem cells and epigenetics might come to play a greater role. In the end there may be more than six hallmarks. The hope is that the number will be finite and reasonably small.

One evening during the meeting I came upon a crowd of scientists streaming into a hotel ballroom exhausted from a day of absorbing and exuding information. Inside lavish buffet tables were placed strategically—roast beef with Oregon blue cheese, roasted chicken breast caprese, miniature crab cakes, Southwest chicken empañadillas. Bartenders at six stations offered copious pours of good wines. It was the annual reception for the MD Anderson Cancer Center. Since Nancy and I had gone there one sad January for a second opinion, the institutional logo had been changed. A slash had been added through the word "Cancer." I wondered what marketing fool had come up with that. It seemed tacky, and from the point of view of so many of cancer's victims cruelly optimistic.

From the Anderson affair the crowds flowed onward into a larger ballroom for more drink and dessert and dancing courtesy of the AACR. A soul band, lit from behind with a blue and red spotlight, was playing an old Smokey Robinson tune as the singer, carrying a wireless microphone, tried to coax people onto the dance floor.

First there were two couples dancing, then half a dozen, and by ten o'clock there were fifty, swirling like a whirlpool and pulling others onto the floor. As I walked back out in the hallway, the rhythm had slowed and the lights had dimmed. The singer was singing "Killing Me Softly." Which is exactly what cancer doesn't do.

## Chapter 10

# The Metabolic Mess

In 1928 in a laboratory at St. Mary's Hospital in London, Alexander Fleming discovered penicillin. He had been growing staphylococcus bacteria on a culture plate, and upon returning from a holiday he noticed that it had been contaminated by a spot of mold. Around the spot were the corpses of dead bacteria. Fleming isolated the fungus and found that he could dilute it a thousandfold and it was still potent enough to kill the microbes. He went on to show that the mold, from the genus *Penicillium,* was also effective against streptococcus, pneumococcus, meningococcus, gonococcus, diphtheria, anthrax—so many killers that can now be rendered harmless with a few shots of antibiotic, letting us live long enough to get cancer.

St. Mary's has since been absorbed as a campus of Imperial College School of Medicine, and I was walking there one afternoon across Hyde Park to see Elio Riboli, director of Imperial's School of Public Health. Riboli's career as an epidemiologist has spanned four decades, leaving him particularly well suited to reflect on the changes in our ideas of what does and does not cause cancer. Chemical carcinogens appeared to be much less of a factor than I had suspected, and the case for or against certain foods was as blurry as ever. Riboli seemed like a man who could help straighten out the confusion.

It was a clear spring day, and as I walked I tried to imagine the gloom of the Industrial Revolution when the air was thick with smoke and coal dust. It was in London in the late 1700s that Percivall Pott drew the connection between exposure to soot and scrotal cancer among chimney sweeps—one of the early observations in mankind's groping toward a theory of cancer. Chimney sweeps were not such jolly characters as the one Dick Van Dyke played in the movie *Mary Poppins*. Boys thin from malnutrition were induced for a few farthings to slither, often naked, through the grimy passages. "The fate of these people seems singularly hard," Pott wrote. "In their early infancy they are most frequently treated with great brutality, and almost starved with cold and hunger; they are thrust up narrow, and sometimes hot chimneys, where they are buried, burned and almost suffocated; and when they get to puberty, become liable to a most noisome, painful, and fatal disease." Treatment involved removal, without anesthetic, of the tumorous part of the scrotum. This had to be done immediately. Once the cancer had spread to a testicle it was usually too late even for castration.

> I have many times made the experiment; but though the sores, after such operation, have, in some instances, healed kindly, and the patients have gone from the hospital seemingly well, yet, in the space of a few months, it has generally happened, that they have returned either with the same disease in the other testicle, or in the glands of the groin, or with such wan complexions, such pale leaden countenances, such a total loss of strength, and such frequent and acute internal pains, as have sufficiently proved a diseased state of some of the viscera, and which have soon been followed by a painful death.

The cause of the cancer was presumably the grinding of soot into abraded skin. Chimney sweeps on the European continent, who wore protective clothing—their outfit resembled a diving suit—didn't get the cancer, and it was unknown in Edinburgh, where chimneys, less angular and narrow than those in London, were usually cleaned from above with a broom and an attached weight. But it was impossible to draw a simple arrow between cause and effect. Even among the London sweeps, the cancer was very rare and might take twenty years to develop. And why did it almost always affect the scrotum—there were a few reports of soot warts on the face—but not other parts of the body exposed to the same scraping application of the carcinogen? There must have been other factors involved. I thought of the experiments in the early twentieth century when a Japanese scientist, Katsusaburo Yamagiwa, induced tumors varying in size "from that of a grain of rice to that of a sparrow's egg" by applying coal tar to rabbits' ears. But it was a painstaking procedure, fraught with failure, and the tumors appeared only after repeated applications of the cancerous grime.

Occupational exposures were also the preoccupation of Bernardino Ramazzini, an Italian physician who wrote *De Morbis Artificum Diatriba (Diseases of Workers)*, published in 1700. He was comprehensive in his interests, studying not only laborers and tradesmen but also apothecaries, singers, laundresses, athletes, farmers, and even "learned men," which included mathematicians and philosophers as well as physicians like himself. All were prone to various afflictions, but the only cancer he mentioned in the book occurred in nuns. Ramazzini noticed that they tended to get more breast cancer than other women. "Every city in Italy," he wrote, "has several religious communities of nuns, and you can seldom find a convent that does not harbor this accursed pest, cancer, within its walls." He attributed this to celibacy and a "mysterious sympathy" between the uterus and breasts, one that also would explain how milk conveniently appears in the mammary glands of women when they become pregnant. "We must certainly believe that the Divine Architect fashioned the uterus and the breasts with some structure, some contrivance that so far escapes us," he wrote. "Perhaps the course of time will reveal it, since the whole domain of Truth has not yet been conquered."

It was not until the twentieth century that scientists began to elaborate the complex system of sex hormones that travel through the bloodstream to distant parts of the body. Among their many roles is coordinating the activity of the uterus and breasts. By forgoing the bearing and nursing of children and experiencing more menstrual cycles, the nuns had unknowingly increased their exposure to their body's own carcinogenic estrogen, accelerating cellular division and raising the odds of mutation.

There was also a benefit to a lifetime of celibacy. A century and a half later another Italian, Domenico Rigoni-Stern, observed that nuns got less cancer of the cervix, foreshadowing the discovery that the principal cause is the human papillomavirus, acquired through sexual intercourse. Chimney soot, sex hormones, in a few cases viruses—there are so many things that can set off a cellular explosion and so many factors that remain to be understood.

Riboli, who earned an MD and a master's degree in public health from the University of Milan in 1980, was part of a venerable line of Italian physicians seeking clues to the vagaries of cancer. From Milan he went to Harvard for another master's in epidemiology. When I arrived at the London campus he was waiting in his office. He is tall and thin, a courtly, soft-spoken man who has taken to heart the evidence that controlling one's weight and exercising provides an edge over both heart disease and cancer. For the next hour and a half, we talked about what he had learned in the course of his epidemiological research. Looking back months later, I was struck again by the whipsaw effect of nutritional science, where what is good for you one day may be bad for you the next, and I wondered: How much can we really control whether or not we get cancer?

By the time Riboli had begun his career it was clear that tobacco smoke was causing an epidemic of lung cancer, and it seemed sensible that other cancers would also be traced to particular chemicals—the industrial contaminants that were being added to the air and water, the preservatives and pesticide residues in food. "The dogma was that cancer *must* be caused by carcinogens," he said. Chemicals, viruses, bacteria—some influence from beyond. But early on there were signs that the hypothesis wasn't holding up. "Despite extensive research for some of the most common cancers—like cancer of the breast, cancer of the colon, cancer of the prostate—no single carcinogen was found to play a meaningful role in humans." Riboli was not saying that cancer-causing agents were having no influence on the population. "People can be exposed to a large number of carcinogens in the air and the water which can and actually do cause cancer. But for as much as fifty or sixty percent of cancer, we didn't have the slightest idea of where it comes from."

Only in a minority of cases could the blame be put squarely on inherited genetic defects. The migrant studies had established that. People moving to new countries, carrying along the same genes, had an increased risk of acquiring, within a generation, the cancers of their hosts, and they often left behind the cancers of their homeland. As Doll and Peto's influential study suggested, the most important factor was human behavior, and a consensus was beginning to form that the likeliest contributor was what we ate.

The first clues came from laboratory experiments. Instead of applying coal tar to lab animals' ears, researchers tried feeding them different amounts and varieties of foods to see how fat—or adipose—they would get. "In a number of experiments no chemical carcinogens were used, but by modulating the diet—by modulating the adiposity—it was shown that you could modulate the frequency of tumors," Riboli said. At first it seemed that an excess of fatty foods was the reason. But further research suggested that it was not so much the fat or other ingredients that were to blame but the total intake of calories—that obesity itself was a primary force in cancer.

Some foods appeared to pose small risks. Diets too rich in salt were associated with stomach cancer and red and processed meat with colon cancer, possibly because of nitrosamines, N-nitroso compounds, and other substances. "There was not a very strong association as there is with smoking and lung cancer, where the effect is gigantic," Riboli said. "We were talking about an increasing risk of one and a half to twofold for some lifestyle habits compared to others." When a risk is very small to begin with, even doubling it leaves a person with very little chance of getting cancer. But spread across populations of millions, the effects could have a significant impact on public health. Investigating that further, however, would require large epidemiological studies, which can be frustratingly difficult to interpret.

"The nineteen-eighties was an extremely challenging period," Riboli recalled. Cancer researchers polarized into two factions. He was reminded of Dante's Florence, where the warring Guelphs divided into the Neri and the Bianchi, the Blacks and the Whites. "We had two parties, one saying it's all environmental carcinogens and the other saying cancer can develop without them. I moved from the carcinogenesis party to the lifestyle party." He became interested not only in the factors that might cause cancer but also in the ones that might prevent it.

During the decade that followed, he helped with an effort, organized by the World Cancer Research Fund and the American Institute for Cancer Research, to review some four thousand studies on nutrition and cancer and see what patterns emerged. In 1997 the groups issued their report, *Food, Nutrition and the Prevention of Cancer: A Global Perspective*—inspiration for the 5 A Day program that was all the rage in the years before Nancy's diagnosis. Based on the best available evidence, fruits and vegetables appeared to have remarkable powers: "Diets containing substantial amounts of a variety of vegetables and fruits may on their own reduce the overall incidence of cancer by over 20 percent." The number-one recommendation was to eat "predominantly plant-based diets" with five or more servings every day. In her widely read column in *The New York Times*, Personal Health, Jane Brody gave her summary of the study's remarkably specific recommendations:

Foods especially rich in cancer-protective chemicals include the onion family, cabbage-family vegetables (including broccoli, cauliflower, bok choy, kale and brussels sprouts), dried beans and peas, tomatoes, deep yellow-orange vegetables and fruits (like sweet potatoes, cantaloupe and winter squash), citrus fruits, blueberries and dried fruits like prunes and raisins.

If only it had turned out to be so easy. A decade later, in 2007, came the disappointing follow-up. Riboli was again a key member of the study. As more and better evidence accumulated, the case for fruits and vegetables was unraveling. There was still "limited" to "probable" evidence that some of these foods might slightly lower the risk of certain cancers. But the authors concluded that "in no case now is the evidence of protection judged to be convincing."

The problem with the earlier report (and to a lesser extent with its sequel) is that the conclusions were based so heavily on retrospective studies, those where you must rely on people to remember in detail what they had eaten years and even decades earlier—the gestation period for many cancers. "If you ask someone who is seventy years old who has colon cancer what his diet was when he was forty-five or fifty, it's a tough call," Riboli said. "For things like smoking or drinking it's more clear-cut. These are things that are very repetitive and stable." Things that you remember. "But how often do you eat carrots? How often do you eat pears? Quantify how many pears, how many strawberries, how many eggs—including all the eggs you don't know about because they are in recipes." Riboli believed better answers lay in prospective studies, the kind that followed a large population of people as they went about their lives. Then those who developed cancer could be compared with those who did not. "We wouldn't need to go to someone who is there in a bed with cancer and ask how often he was consuming salads," Riboli said. "We collect information from people who are living their normal life."

While the projects by the World Cancer Research Fund were under way, Riboli had been pushing to form EPIC, the European Prospective Investigation into Cancer and Nutrition. During the 1990s, researchers had begun monitoring the health of 520,000 people in ten countries. Blood samples were drawn periodically and preserved with liquid nitrogen. Heights, weights, and medical history were recorded. Information was collected on diet and physical activity. As, year by year, the database grew, investigators at various universities and government agencies began looking for correlations.

A few early results had found their way into the 2007 report, helping to tip the balance away from a preoccupation with fruits and vegetables. Since then more surprises have appeared. By the time I talked to Riboli, about 63,000 of the half-million people in the study had cancer. There was now only the slightest evidence that eating a lot of fruits and vegetables had made much difference. They did not obviously reduce the overall risk of getting cancer or even of specific cancers like those of the breast, prostate, kidneys, and pancreas. There were suggestions of a small protective effect, especially among smokers, for cancer of the lungs, mouth, pharynx, larynx, and esophagus. But it was too early to make more than tentative conjectures. Besides smoking, a risk factor for many of these cancers was heavy drinking, and people who smoke and drink excessively have been shown, as might be expected, to be less likely to eat fruits and vegetables. A preliminary study found that these foods possibly played a small part in reducing cases of colon cancer but that too remains in dispute.

In an editorial for the *Journal of the National Cancer Institute,* Walter C. Willett, a prominent nutritionist (he was the head of the influential Nurses' Health Study on diet and lifestyle) and one of Riboli's longtime colleagues, concluded that researchers had been "overly optimistic" and that the EPIC findings only added to the evidence "that any association of intake of fruits and vegetables with risk of cancer is weak at best." It had become clear with Doll and Peto that synthetic carcinogens were not the smoking gun, and now it appeared that fruits and vegetables were not a magic bullet.

Diet was not irrelevant. It was EPIC researchers who estimated that for a 50-year-old who ate a lot of red and processed meat (160 grams, or more than a third of a pound a day) the ten-year risk of getting colorectal cancer was 1.71 percent—0.43 percent higher than for someone who ate less than 20 grams. A third of a pound a day is a lot of hamburgers and hotdogs, and again there are complications to bear in mind. The study made adjustments for smoking, drinking, and other confounding factors. But there may be something else about the behavior of carnivores that skewed the results, and other studies have come to conflicting conclusions. There will always be uncertainties with observational epidemiology and the inevitable question of what is cause and what is effect. Getting closer to the answers would require very large randomized trials in which one population would faithfully eat more of some food and the other would eat less. After twenty or thirty years of draconian enforcement, maybe you could say with some confidence whether there was a difference in the risk of cancer. The data EPIC hopes to collect in the coming decades might be the next best thing.

Moving beyond purely culinary issues, EPIC has strengthened the case against obesity. One study found that older women who had gained 15 to 20 kilograms, or roughly 40 pounds, since they were twenty had an increased breast cancer risk of 50 percent. As in the old animal experiments, fatness itself, whatever its cause, appeared to be the driving force. Along with lack of exercise, it may account for as much as 25 percent of cancer, with dietary specifics falling to as little as 5 percent. This is the message emerging from decades of nutritional and medical research: Understanding cancer lies less in the foods we eat than in how the body stores and uses energy.

At the center of this metabolic puzzle is the hormone insulin. As we eat and our level of glucose (blood sugar) rises, it is insulin, secreted by the pancreas, that signals our cells to burn the fuel directly and to store the excess as glycogen (starch) or body fat. As blood sugar falls, the cells draw on their reserves by converting glycogen back into glucose. When still more energy is needed, fat cells release their long-term supplies. Sometimes, however, something goes wrong. The body produces too little insulin or becomes numbed to its effects. When the latter occurs the pancreas responds by producing more insulin. The cells react by becoming even more resistant and so more insulin is secreted. This pathological spiral—a condition called metabolic syndrome—is involved in chronic conditions like hypertension, cardiovascular disease, diabetes, and obesity. It also plays a role in cancer. The reasons are complex. Insulin and closely related hormones called IGFs (insulin-like growth factors) can stimulate a cancer cell, feeding the expansion of tumors and even encouraging angiogenesis. Insulin is also involved in regulating sex hormones. Moreover, a rise in insulin accelerates the accumulation of body fat, and fat cells synthesize estrogen. Insulin, estrogen, obesity, cancer—all are tied into the same metabolic knot.

It makes sense that connections like these would have evolved. A woman must be well nourished in order to produce healthy babies. In times of famine, there is no excess energy to store, and the metabolic machinery reacts by lowering the availability of estrogen. It is not a good time to conceive. As more food becomes available, fat accumulates—energy the mother will need during pregnancy and nursing—and more estrogen is released, stimulating ovulation and, after conception, the production of breast milk. Here is the basis of the "mysterious sympathy" Ramazzini wondered about more than three centuries ago. But in a civilization where food becomes abundant, and overly so, the sympathy is upset. The age of menarche decreases, adding to the number of estrogen cycles and raising the risk of breast cancer. Increased nutrition may also unleash the hormones that produce greater body height—another risk factor for cancer. "What this shows," Riboli said, "is how something which is just a modulation of a normal physiological process—which remains normal and doesn't cause any disease—has a major impact later in life with cancer. This is not chemical or physical or viral carcinogenesis. It is metabolic carcinogenesis." The ancient idea of cancer as a disposition of the whole body has returned in a more sophisticated form.

The amount of fat in storage also affects the immune system in ways that might promote malignancy. In addition to fat cells, fatty tissue contains gobs of macrophages—cells that flock to infected trouble spots to ingest invaders and that can also be diverted to aid in a cancerous attack. And the fat cells themselves secrete other agents that promote inflammation—a healing mechanism that involves the rapid creation of new tissues. There is a thin line between that and tumorous growth. More than a century ago Rudolf Virchow suggested that chronic inflammation, with its power to accelerate cellular proliferation, was a cause of cancer. (That might explain why aspirin and other anti-inflammatory drugs appear in some studies to lower cancer risk.) Obesity has been described as a kind of "low grade chronic inflammatory state" and tumors as "wounds that do not heal." Chemokines, integrins, proteases . . . neutrophils, monocytes, eosinophils—there is so much invisible apparatus behind the crude feel of a throbbing joint or a hot, pus-filled wound. Inflammation has also been tied to metabolic syndrome and diabetes. Cancer, obesity, diabetes—the strength of these connections is hinted at in studies of grossly overweight people who undergo gastric bypass surgery in a last-ditch attempt for relief. As their body mass decreases their diabetes recedes, and there is evidence that they get less cancer.

The deeper you look the more convoluted this all becomes. Cortisol, the stress hormone, and melatonin, which regulates sleep, are also hooked into the metabolic loops involving energy, estrogen flux, and inflammation. Epidemiological studies have suggested that women who work at night may have a higher risk of breast cancer. Considering that and other evidence of the effect of sunlight and sleep cycles on the body, the World Health Organization added "shiftwork that involves circadian disruption" to its list of probable carcinogens—one more avenue that may warrant exploration. All these phenomena are joined at the cellular roots and understanding cancer will require sorting them all out. The overall incidence of cancer has leveled off in recent decades. Are our bodies learning to adjust to the new rhythms? We can never know for sure how cancer rates in the twenty-first century compare with those hundreds of years past. If over the long run there has been an increase, then part of the story could be the modern changes shaking our metabolic core.

By the time I caught up with Riboli, he and his colleagues were talking less about broccoli, cauliflower, bok choy, kale, and brussels sprouts and more about the body's energy balance and how the fulcrum has shifted since ancient times. I'd read the debates about the so-called paleo diet—was it richer in fruits and vegetables or in meat and fat? In any case it was low in refined carbohydrates and sugar—energy-packed foods that hit the blood so quickly, causing spikes in insulin and potentially disrupting so many biochemical cascades. Toward the end of our interview Riboli pulled from his bookcase a binder of charts. "At the end of 1800 the usual consumption of sugar in most European countries was two to three kilograms per person per year," he said. "Now it is between fifty and sixty kilograms." I pictured a hundred-pound pile of sugar and eating it over the course of twelve months. I was reminded of the journalist Gary Taubes, who argues that carbs and sugar, rather than dietary fat and overeating, drive the modern obesity epidemic and the damage it causes, including cancer, by skewing how the body uses energy.

Riboli and his colleagues suspect that all energy-rich foods are a problem. Although they are high in calories they can leave us unsated and wanting more. "If I go and buy a burger or a sandwich, most often it contains between five hundred fifty and six hundred kilocalories," he said. "If I prepare a nice pasta dish, Italian style—with some sauce, pimento, some vegetables—I barely reach five hundred kilocalories. But I have something so voluminous that I feel full. I eat a sandwich and have the impression that I haven't eaten anything, but I've had more kilocalories—more energy." That empty feeling might spur the desire for a candy bar. Maybe that is reason enough to eat more fruits, vegetables, and fiber. They fill your stomach, reducing your energy intake and therefore your insulin load.

The other side of the energy equation is physical exercise, and in modern times people are able to lead more sedentary lives. "You and

I are having a very pleasant conversation sitting here," Riboli said. "At another time in another place we might be having this conversation walking in a field. We are moving less and eating more." Exercise is not, however, a simple matter of burning off pounds. Exertion makes you hungry and you may respond by consuming at least as many calories as you expend. More important may be the effects of exercise on keeping insulin and other hormones under control. Lower your weight and exercise more. "Twenty years ago these were just ideas," Riboli said. Now EPIC is seeking scientific support. The work is only beginning. An official statement from EPIC promises to explore the complex interactions between genetic, metabolic, hormonal, inflammatory, and dietary factors. More knots to untangle.

I told Riboli I was feeling even better about having walked to his office all the way across Hyde Park. He laughed and as I put away my notebook he took me on a brisk walk down the hall, out of the building, and beyond the gate of the hospital grounds, until we were standing on the side of Praed Street. He pointed up to a window in the old hospital building, the one to Alexander Fleming's office. He told me a story that has become part of the legend—how Fleming had accidentally left the window open, allowing the spores of penicillin fungus to contaminate the agar plate. That detail may be apocryphal, but it is an encouraging reminder that a great medical discovery can come suddenly through an act of serendipity.

As I walked toward the Tube station—I'd exercised enough for the day—I thought of how it can never be so easy with cancer. The infectious diseases we have defeated were each caused by a single agent—an identifiable enemy that could be killed or vaccinated against. With cancer we would have to seize control of a whole slew of factors, including the mishmash of symptoms arising from imbalances in energy metabolism. And the biggest risks will always lie beyond our grip: old age and entropy. Cancer is not a disease. It is a phenomenon.

What left me feeling more optimistic is what EPIC might find in the future. In coming years as more people in the study come down with cancer, researchers will be able to analyze their blood in minute detail to see what it was like years or even decades before they got sick. With technologies like nuclear magnetic resonance, they will be able to scrutinize thousands of blood chemicals, looking for signs that might portend the later onset of cancer. This is a very different way of doing medical research. A scientist traditionally begins by posing a hypothesis—based on an observation or a statistical study or a simple hunch. Maybe a high level of a vitamin increases or lowers the risk of some cancer. Then you go looking for evidence. With resources like those at EPIC, connections may emerge that no single mind would have come to suspect. The result could be reliable tests that give early warning for a malignancy the way high cholesterol warns of heart disease. Maybe by then there will be something we can do about it.

## Chapter 11

# Gambling with Radiation

One surefire carcinogen Riboli and I didn't talk about is radioactivity. Here the mechanism is straightforward: The unstable nucleus of an element like radium shoots out particles and rays with so much energy that they can tear through molecules, break chemical bonds, and wreak all kinds of cellular hell. Emanations this forceful are called ionizing radiation (atoms stripped of electrons are ions). If the radioactive particles don't strike a gene head-on, inducing a mutation, they might leave a wake of corrosive free radicals in the cell's cytoplasm—a condition called oxidative stress than can damage the genome indirectly. Shifting into panic mode, the mangled cell might send signals to neighboring cells, inducing more stress and genomic shock. Most of the exposure we receive from this carcinogen comes from natural sources. The greatest contributor is said to be radon rising from the soil below.

Ever since I had my house tested for the gas two decades ago, registering a modest amount, I had paid little attention to the warnings. Radon, like carbon monoxide, is an invisible, odorless, silent killer— albeit one that works slowly as mutations mount year by year. Of the approximately 160,000 lung cancer deaths each year in the United

States, the Environmental Protection Agency has said that 21,000, or 13.4 percent, may be radon related. What you don't often hear is that for about 90 percent of those deaths smoking is also a factor. Through all the years of my life, I had smoked a grand total of maybe ten cigarettes—and none during the last twenty-five years. Still, as I began to learn more about cancer, I felt a need to conduct another radon test—this time in a room where I had recently been sitting for weeks writing this book.

It had been an unusually cold winter in Santa Fe. Access to my second-story office requires traversing an outdoor staircase. It's an easy and picturesque commute but sometimes it involves shoveling snow. For that and other reasons I had taken to working downstairs in a room that was built, like many in old Santa Fe, over a dirt crawl space. Two walls of the room are about six feet below ground level and built from adobe bricks molded from the same dirt that lay beneath the floor. For weeks the weather outside had been too cold for opening windows, and I had latched shut a door between the office and the hallway to hold in heat. The conditions, in other words, were likely to result in stagnant air and maximum readings of radon gas.

I ordered a test kit, placed it on the desk, and forty-eight hours later mailed it to the laboratory named on the instruction sheet. This time the results that came back were more than quadruple what they had been before: 22.8 picocuries per liter. The EPA's scale, correlating radon levels with risk, topped off at 20, and remedial action was recommended at just 4 picocuries per liter. A curie is approximately the amount of radiation produced by a gram of radium, so a picocurie is one-trillionth of that: 2.2 nuclear disintegrations per minute. As radon rapidly decomposes, it shoots out alpha particles (clusters of two neutrons and two protons) and breaks down into smaller elements, which float through the air emitting alpha particles of their own. They don't travel far—alpha rays can be stopped with a sheet of paper—but because of their massiveness they deliver a heavy blow. The radon gas itself is readily expelled from the lungs, but the daughter particles, inhaled with every breath, can stick in the wetness and irradiate cells. Every minute in every liter of that stagnant air, fifty of these submicroscopic explosions were occurring. The EPA chart that came with the test kit informed me that if one thousand people who have never been smokers are exposed to 20 picocuries per liter throughout their entire lives, thirty-six of them would be likely to get lung cancer. Another way of saying it is that the lifetime risk is 3.6 percent. (For smokers exposed to that much radon the odds are seven times greater.)

As I thought about these numbers, I began to feel a tightness in my chest. I imagined my lungs heavy with a miasma of cold, radioactive air. Compared with the astronomical amount of atoms in one breath of air, the fifty radioactive events occurring every minute is a vanishingly tiny proportion. And only a fraction of the shrapnel, these alpha particles, would strike lung tissue and cause genetic mutations. Most mutations, I reminded myself, are harmless. Our DNA is mutating all of the time. Cells have evolved mechanisms to repair broken DNA or to destroy themselves if the damage is too great. Of all the mutations that occur in a genome only certain combinations might trigger a cancer, and only if many other things go wrong. But for all of those reassurances there still was a palpable risk.

The test had been done under such airtight conditions that the reading was bound to be abnormally high. Half a year later, in warmer weather, I measured again. This time I placed the detector in the bedroom (where Nancy and I had slept for seventeen years). I opened and closed doors and windows according to my usual routine. The measurement this time, closer to normal conditions, was much lower—7.8 picocuries. A third reading, in the hottest part of summer, when fans were circulating air through the house, came in at just 0.8 picocuries—way below the national mean. The average of my three readings was 10.5 picocuries (a risk of 1.8 percent). My odds were looking better and I wondered if I could lower them a little more.

The EPA numbers are based on the assumption that people

spend, on average, 70 percent of their time at home—nearly seventeen hours a day. That would be high for someone commuting to a full-time job. I work at home but most often I am upstairs, where my exposure would presumably be much less. Radon comes from the earth and is eight times heavier than air. With no interior staircase or forced-air heating, I felt safe in my office aerie. When I am downstairs I am often in parts of the house where radon levels are also probably lower. (Maybe I will buy more test kits.) To allow for all of that, I cut my estimated exposure—reducing it by one-fourth seemed reasonable—and then I cut it again. I've lived in the house for only about a third of my life. Dividing by three brought the level down to 2.6 picocuries—below the EPA "action level"—and my risk to about 0.3 percent. The chance of a nonsmoker getting lung cancer sometime in life is usually put at around 1 percent or less. If so, then living in this comfortable old house might have raised my odds to something like 1.3 percent, from a minuscule risk to a somewhat less minuscule one. But I guess that is a self-centered view. Spread across the population that would account for a lot of cancer.

My calculations were rough. If I wanted to estimate more precisely, I would have to consider every other place I have lived. I had a basement bedroom when I was a child, but I'd lived on the fourth floor of a row house in Brooklyn and the eighteenth floor of a highrise in Manhattan. It would be possible in theory to calculate long-term exposure with a laboratory analysis of my eyeglasses. When alpha particles strike carbonate plastic lenses they leave tracks—memories of radiation exposure. The tracks—there are typically thousands per square centimeter—can be translated into radon readings. There is also a method using ordinary household glass. Radon decay products are deposited on mirrors, picture frames, and cabinet windows and can become incorporated into the glass. By measuring the amount that has accumulated and considering other variables, epidemiologists can estimate how much radon people have been exposed to over many years—not just in their current homes but for as long as they have owned the objects.

As I thought about all the microscopic wallops I might have

incurred, I wondered where the EPA had gotten its figures in the first place—so many picocuries per liter corresponding to so many lung cancer deaths. It's not like you can lock a thousand people in a basement and then wait for some of them to get cancer. The story began in the 1970s when houses in Grand Junction, Colorado, built on top of tailings salvaged from uranium mines, were found to have elevated levels of radon. At great expense the radioactive fill was removed and replaced, but the radon readings remained high. Then came a much reported incident with a construction engineer named Stanley Watras. He was working in 1984 at a nuclear power plant in Pennsylvania. As the plant neared completion, radiation alarms were installed, and they sounded whenever Watras passed by. The reactors, however, were not yet operating and there was no fissionable material in the plant. The source of the contamination turned out to be his house, which measured as high as 2,700 picocuries. You didn't need to build on uranium tailings to have radioactive air. Homes across the country were found to register positive for radon, and it was coming from the natural soil. Radon has been with us from the start.

In an attempt to gauge how threatening the exposure really was, epidemiologists began conducting case control studies, comparing radon levels for people who had contracted lung cancer with people who had not. Early results were inconclusive—some detected a small effect and others did not. A study in Winnipeg, which had the highest radon levels of eighteen cities in Canada, found no influence on lung cancer. Other researchers compared the average radon levels of different geographical areas. Again no association was found. A nationwide survey reported a negative correlation, as though breathing radon somehow provided protection. Or else the study was flawed. Some critics suspected that the results were skewed by an inverse connection between smoking and the amount of radon measured in homes. Perhaps cigarette smoke interfered with the radon monitors, or smokers were more likely to occupy older, draftier houses or to open more windows.

Getting better numbers would require either very large popula-

tions or very high radon levels—the hundreds to thousands of picocuries per liter that can be found in underground mines. Looking for
answers, researchers studied lung cancer rates among uranium miners in Colorado, New Mexico, France, the Czech Republic, Canada
(an area on the shore of Great Bear Lake had the evocative name
Port Radium), and Australia (Radium Hill). They studied miners
of other ores in Canada, China, and Sweden—altogether 68,000
men. Of those, 2,700 had died from lung cancer. That is about 4
percent. There were confounding factors to consider. Most of the
miners were believed to be smokers but the data on how long or how
often they had smoked was sparse or nonexistent. Miners are also
exposed to diesel fumes, silica, and other dust, which might have
synergistic effects. Laborers breathe harder than someone cooking
dinner or reading a book in bed.

Doing their best to adjust for these complications, a committee
of the National Research Council began analyzing the numbers and
quantifying the relationship between radon and lung cancer. They
assumed that it must be linear—that one-tenth the exposure leads to
one-tenth the risk. Not all toxicologists believe that is true, proposing instead that there is a threshold below which radiation causes no
damage. But the mainstream view is that even the smallest amounts
are potentially harmful. With marathon feats of statistical calculation, the numbers for the miners were adjusted downward to estimate the risk from the far lower exposures found in homes. That
was the basis for the chart distributed by the EPA and included in
my test kit.

Some critics thought that extrapolating from miners to suburban
neighborhoods was too big a leap. But in recent years the estimates
have been supported by more extensive research on households. The
most ambitious study was carried out in Iowa. The state has the
highest average radon levels in the country. Women were chosen as
subjects because they were more likely to spend time at home. To
qualify they had to have occupied the same house for at least the
past two decades. Radon detectors were placed in several locations

in each house and readings were taken over the course of a year. Through questionnaires the researchers estimated the percentage of time the women spent in various rooms or other buildings—or outdoors, where average radon levels were also measured. When the women had been on vacations or business trips it was assumed that they received the average exposure for the United States. Allowances were made for occupational exposure, smoking (passive included), and other factors. In the end, it was concluded that someone living for fifteen years in a house with an average radon level of 4 picocuries per liter might have an "excess risk" of about 0.5. The age-adjusted incidence of lung cancer (for smokers and nonsmokers combined) is about 62 cases per 100,000 men and women per year. All things being equal, that would increase by half to 93 cases—31 more people suffering the horror of what is almost always a fatal condition.

No single study can draw firm conclusions. The sample size is too small. But statisticians have gone on to amalgamate the data, producing what is called a pooled analysis. It's tricky work. Research is conducted on different populations according to different methodologies. In combining the numbers these discrepancies must be accounted for. Three of the analyses—in Europe, North America, and China—found similar results to those derived from the experience with the miners, and most radon researchers now consider the matter clinched.

But epidemiology is never a closed book. As I was anxiously poring over the radon literature I learned about a controversial hypothesis called hormesis, which holds that small doses of radiation are not just harmless but beneficial. We evolved in a world bathed in radiation, the argument goes, and have adapted to all but the most egregious assaults. A Johns Hopkins researcher recently concluded that levels of radon as high as 6.8 picocuries per liter may actually lower lung cancer risk. While the alpha particles are causing potentially carcinogenic mutations, low-level x-ray, gamma, and beta radiation may be activating epigenetic circuitry involved with DNA repair and apoptosis and enhancing the immune response. If that is true then reducing exposure to the EPA's recommended action level might actually increase lung cancer risk. But that remains a maverick view. Considering the evidence on balance, I decided to start keeping a window cracked open when I work downstairs, even on cold days in winter. Just in case it matters.

Not even the radiation from nuclear blasts, accidental or deliberate, has caused nearly as much cancer as most people think. Fifty workers were killed almost immediately by the estimated 100 million curies unleashed by the calamity at the Chernobyl nuclear power plant in 1986. A huge wave of cancer was expected to follow. But almost two decades later a United Nations study group lowered its estimate of the excess: 4,000 deaths among the 600,000 people (workers, evacuees, and nearby residents) who received the highest exposures, or less than 1 percent. There was an increase in thyroid cancer among people exposed as children, but the biggest public health problem, the report concluded, has been psychological. "People have developed a paralyzing fatalism because they think they are at much higher risk than they are, so that leads to things like drug and alcohol use, and unprotected sex and unemployment," a researcher told *The New York Times*. The government of Ukraine recently opened the Chernobyl site to tourism, and ecologists have found that the absence of humans has turned the area into a mecca for wildlife.

The nuclear warheads dropped on Hiroshima and Nagasaki in 1945 killed at least 150,000 people—either immediately from the impact or within months from injuries and radiation poisoning. Since then scientists have been monitoring the health of approximately 90,000 survivors. They estimate that radiation from the explosions has led to 527 excess deaths from solid cancers and 103 from leukemias.

Tsutomu Yamaguchi survived both blasts. Visiting Hiroshima on a business trip, he was close enough to ground zero to suffer severe burns and a ruptured eardrum. After spending a night in a shelter, he returned home to Nagasaki in time for the second blast. He died in 2010 at age ninety-three. The cause was stomach cancer. It's impossible to know how big a factor radiation played in the death of the old man, who had outlived so many others. Maybe the crowning blow was a diet of salted fish.

It was leukemia that killed Marie Curie, the discoverer of radium (radon's mother), at age sixty-six—"cancer in a molten, liquid form," as Siddhartha Mukherjee memorably called it in *The Emperor of All Maladies*. When she was exhumed in 1995 for the honor of being reburied with Pierre in the Panthéon, French officials worried that her body would be dangerously radioactive. The three black notebooks describing her celebrated experiments are kept in a lead box at the Bibliothèque Nationale in Paris, and people who want to read them must sign a waiver acknowledging the risk. When her grave was opened her remains were found enclosed in a wooden coffin inside a lead coffin, which was inside another wooden coffin. Emanating from inside was 9.7 picocuries—almost twenty times less than the maximum considered safe for the public by the French government. Madame Curie was only half as hot as the air on that winter day in my office.

With a half-life measured in centuries, the radium she had absorbed during her career would not have appreciably diminished since her death. France's *Office de Protection Contre les Rayonnements Ionisants* therefore concluded that it probably wasn't radium that killed her. A more likely cause of her cancer, they suggested, was the x-ray equipment she and her daughter, Irène Joliot-Curie, operated as medical volunteers in World War I. Irène, who won a Nobel Prize for her own work on radioactive elements, also died from leukemia. She was fifty-eight.

For Pierre, death came early, at age forty-six, when he was run over on a Paris street by a horse-drawn carriage. We don't know what kind of damage radium might have done to his cells. He and Marie had both been too ill to travel to Stockholm to accept their

Nobel Prize. Whether it was from radiation poisoning or physical exhaustion—extracting a gram of radium from a ton of pitchblende was like factory work—is unknown. Two years later they made the journey. In his Nobel lecture (also delivered on Marie's behalf), Pierre described an experiment he had done on himself: "If one leaves a wooden or cardboard box containing a small glass ampulla with several centigrams of a radium salt in one's pocket for a few hours, one will feel absolutely nothing. But fifteen days afterwards a redness will appear on the epidermis, and then a sore which will be very difficult to heal. A more prolonged action could lead to paralysis and death." This destructiveness, he noted, had its uses. Radium was already being used to burn away tumors. So were x-rays, since just after their discovery in 1895. Long before it was established as a cause of cancer, radiation was used as a cure.

Before Nancy's chemo had ended her doctors began discussing the next stage of her treatment and what kind of particles they should use to irradiate her. Alpha particles are too massive and damaging to beam directly at the body. Beta rays, consisting of streams of electrons, are a gentler radiation. The lightweight particles penetrate a little deeper than alphas—it takes a sheet of aluminum to stop them—but they deliver less punch. They are often chosen to treat skin cancers, sparing what lies below. X-rays and gamma rays have the long reach needed for deeper cancers. Their wavelengths are so tiny that they can pass through many layers of tissue before striking their target. But their fuzzy edges make it harder to avoid harming nearby cells. Protons, which are 1,800 times heavier than electrons but smaller than alpha particles, can deliver large amounts of energy with less mess.

Instead of beaming rays from outside, oncologists might decide instead on brachytherapy: small capsules of radioactive isotopes inserted in or near a tumor. For some cancers, radioisotopes are injected into the bloodstream. Radioactive iodine, for example, will concentrate in the thyroid and attack malignancies there. A targeted drug called Alpharadin delivers radium directly to metastatic bone cancer cells. Whatever the method, the rationale is the same as with chemotherapy: Rapidly dividing cancer cells will succumb more quickly to the poison than healthy cells, and they will be less able to repair themselves.

Both Nancy's surgeon and oncologist agreed that her left and right groin, where the lymph nodes had bulged with carcinoma, should be treated with beta radiation. In the right groin, the cancer had encroached into the epidermal layers, and beams of electrons would impinge just deep enough to reach any cells that the chemo had missed. The doctors disagreed, however, on whether they should also irradiate her entire pelvis with x-rays. The surgeon thought the risks were unwarranted. Radiation can leave internal scars that cause bowel obstructions and it can hurt other organs. Damage to the lymphatic system can bring on lymphedema, an accumulation of lymphatic fluid that can cause chronic swelling of the torso and limbs. Very rarely the mutations induced by radiation will trigger another cancer decades later. There were so many trade-offs to consider.

Certain that he had excised every bit of compromised tissue, the surgeon thought pelvic radiation would be dangerously redundant— that the weeks of chemo followed by superficial beta rays should be insurance enough against escaping metastases. Using more radiation now, when it might not be absolutely necessary, would limit the options if there was a recurrence later on. Both chemo and radiation destroy bone marrow, weakening the body's ability to withstand further therapeutic assaults. "Save your bone marrow for future battles," another doctor advised. Nancy's oncologist was having none of this. He thought that hubris was clouding the surgeon's judgment. So aggressive a cancer in so young and healthy a woman called for an extreme counterattack. Forgoing pelvic radiation, he told Nancy, would be gambling with her life. There was no right answer. The experts at MD Anderson also recommended whole pelvic and that is the course we chose.

Firing rays at cancer cells sounds like a shotgun attack. But the planning and precision is impressive. Medical scanners—CT, MRI, PET—map the tumor and surrounding organs in three dimensions. In aiming the beam, pathways and angles are chosen that avoid the most vulnerable organs. Dosages are calculated meticulously—some organs are more sensitive to radiation than others and so are some tumors. Treatments are scheduled so smaller doses can be spread over days and weeks, gradually enough for the healthy cells to repair or replace themselves but not so gradually that the cancer regains the upper hand. Computer-guided robot arms can deliver graded doses to different parts of a tumor. To reduce the amount of radiation passing through healthy tissue, beams can be pointed from several directions, each weak in itself, converging to deliver the maximum jolt.

For all the care and calculation, damage is unavoidable—the fatigue, the burning skin, the tingling nerves, the diarrhea. Radiation shines through the bowel, creating a sunburn inside. Bulky food worsens the condition, and Nancy was advised to follow a low-residue diet, avoiding high-fiber foods: whole grain breads, coarse-grained cereals, fresh fruits, raw vegetables, wild or brown rice. Also to be avoided were strongly flavored vegetables—broccoli, brussels sprouts, cauliflower—all these things that under other circumstances are supposed to be good for you. All these foods she loved. Chile and other spicy fare, popcorn—all were forbidden. She grew accustomed instead to the taste of Imodium.

Looking back years later through the big ring binder where she kept the papers from this horrible time, I was struck by a couple of absurdities. Among the research papers weighing the dangers and advantages of pelvic radiation and the waivers acknowledging the near- and long-term side effects was a disclaimer: In preparing the patient for treatment ink marks might be made on the body. Nancy had to sign a release recognizing that the ink might rub off on her clothes. She was also advised to avoid getting pregnant.

During the chemo sessions I could sit with her in the light-filled lounge with those beautiful mountain views. For the radiation she was taken into a lead-lined room. Alone with the robot deftly swinging its arm and zapping its preprogrammed targets, she felt like she was in sickbay on the Starship *Enterprise*. She would try to visualize the rays killing the cancer cells and sparing the rest. My strongest memory from that time is the day I drove her to her first treatment. As we approached, she fought back tears. She had been through so much already and I rarely saw her cry. "I can't believe what they're doing to my poor body," she said. And, like so many times, I had to suppress the guilt. I told myself again that her cancer was not known to be estrogen related, that my not wanting to have children was unlikely to be the cause. But who could really know? And what about the stress I had caused her—the blasts of cortisol skewing insulin skewing the metabolic balance? Was there some slight chance—yet to be enshrined in the literature—that radon was a factor? I imagined the gas seeping into pores and orifices. It is part of the curse of being human: this idea that you get cancer because you did something wrong or someone—something—did it to you. For Nancy no cause was ever identified. The best that could be said was that she was a victim of randomness. But randomness can be complexity too deep to understand.

It was during this time that we drove on a Saturday afternoon to the campus of the New Mexico School for the Deaf where the American Cancer Society was holding its Relay for Life. People with cancer are no longer called patients or victims but survivors, and they walked proudly around the track wearing blue T-shirts with a big star and the word "HOPE" in capital letters. (Nancy had another T-shirt at home that said "Not Dead Yet.") I've saved five pictures from then. She is wearing black shorts or a midlength skirt—I can't tell for certain—and I see that her right leg is already swollen with lymphedema. We were assured that it was probably a temporary side effect of the surgery—from damaged lymph vessels—and aggravated perhaps by the treatment. But the swelling never went away. It wasn't such a bad compromise, she would say, for being alive.

Her plan had been to rip off her hat during the procession, bearing her bald head in celebration of having endured the surgery, the chemo, and the first sessions of radiation. But the moment never seemed right. The most memorable part of the day came as the participants walked one by one to the stage, where they briefly introduced themselves, and the first lady of New Mexico bestowed them each with a gold medal and purple ribbon. "I am a cancer survivor," the first woman said, and then the next and the next. I thought of how we've come to sugarcoat our afflictions. Deaf becomes "hard of hearing" becomes "hearing impaired"—and then loops back full circle with the embrace of Deaf Community and even Deaf Culture. Now there is a cancer culture, and whether you had a harmless in situ carcinoma removed with a simple lumpectomy or are fighting the terminal stages of metastatic melanoma, you are called a survivor. In the first case there was nothing to survive. In the second case there will be no survival. The word has been all but stripped of meaning. My thoughts were interrupted when a tall, thin woman with a chemo scarf grabbed the microphone and proclaimed: "I am a *second-time* cancer survivor." Was that really something to celebrate? That the cancer had come back again.

## Chapter 12

# The Immortal Demon

On the early morning flight from Albuquerque to Boston, the captain was wearing a pink tie, and a pink kerchief was peeking from the pocket of his uniform. The flight attendants were similarly dressed, with pink shirts and aprons. It was National Breast Cancer Awareness Month, and when the plane was in the sky one of the attendants enthusiastically announced that the airline was selling pink lemonade and pink martinis—this on a flight departing at 6 a.m. The proceeds would go for "curing" breast cancer.

No more than a hundred years ago cancer was a word spoken only in whispers lest the illness be stirred from its slumber. One might die of "heart failure" or "cachexia," a latinized way of saying that, eaten by cancer, a loved one had wasted away. Though the fear has not disappeared, "cancer" is no longer the unutterable word. The cheerfulness with which the subject has been embraced and shouted is almost macabre. A cosmetics company was advertising "Kisses for the Cure." Buy a lipstick and a small donation would be made to the fight. "Pucker up and Kiss Breast Cancer Goodbye."

As I paged through the airline magazine, I thought of the Stand Up to Cancer telethon I'd watched a few weeks earlier, with singing, laughing, and sometimes somber celebrities vowing to "eradicate" cancer of all kinds. Not control it or reduce it or treat its occurrence more effectively. "Someday no child will die from cancer," a buoyant teenaged actress promised. Not a single one. "We must beat it back and beat it out of existence," said Stevie Wonder, hunched over a piano. His first wife had died from cancer, and many of the other stars had also been closely struck. *"Cancer doesn't care that you've won the Olympic gold medal. Cancer doesn't care if you're beautiful or brilliant or just starting college. . . ."* One by one in their "Cancer Survivor" T-shirts, the idols and their idolizers took the stage. *"Cancer doesn't care if you have your whole life in front of you. . . . Cancer doesn't care that you have young children who need their mother. . . . Cancer doesn't care that it just took your father. . . . It just doesn't care."* A ticker tape message scrolled across the bottom of the TV screen: "Cancer Doesn't Discriminate." But it does. Against the elderly, the obese, the poor. Demographically the young, beautiful people on the show were exceptions. But who could resist their good hearts and cheer? "The stars are taking your calls." And so the telephones rang, the pledges poured in. At the end of the show a procession of scientists filed across the stage to a rousing chorus of "You've got to *stand up, stand up, stand up* to cancer. . . ." Altogether more than $80 million was raised that night.

Stand Up to Cancer is a respected organization reputed to funnel almost all of the money it collects to research. But I wondered if the viewers, as well as the performers, had been left with false hopes. The donations, it was said, would go to "dream teams" of scientists cooperating on a cure instead of competing for recognition and grant money—as if only greed and egos stood in the way of understanding the most complex of medical phenomena. Comparisons were made to Jonas Salk and the March of Dimes, yet polio had been a vastly simpler problem—a disease with a single cause that could be isolated and vaccinated against.

Understanding cancer will require no less than understanding the deepest workings of the human cell. One performer invoked the fight against slavery and the triumphs of the civil rights movement. "What if no one stood up for freedom on the Underground Railroad . . . if no one was standing up for injustice at a bridge in Selma?" Cancer was something to demonstrate against or to oppose with a sit-down strike. These didn't seem like people who were apt to engage in mass acts of civil disobedience like those of ACT UP, the AIDS Coalition to Unleash Power, whose influence lay in its obnoxiousness. Two decades ago, ACT UP demonstrated against the National Institutes of Health and shut down the Food and Drug Administration for a day, demanding more research money and affordable treatments. One way or another more attention became focused on the problem. Now AIDS can be managed as a chronic disease. But not even HIV is as convoluted as cancer.

Descending toward Boston, the plane provided a bird's-eye view of what vies with MD Anderson as the most powerful cancer center in the world. On one side of the river were Dana-Farber, Beth Israel Deaconess, and Massachusetts General Hospital. On the other side were the Whitehead Institute, the Broad Institute, and the campuses of Harvard and MIT. With their petri dishes, gas chromatographs, gene sequencers, and electron microscopes, researchers in these few square miles produce a staggering amount of knowledge about the intricate connections inside a human cell and how they can come undone. For all the horror it causes, cancer is a fascinating intellectual problem—a window into understanding life. But only very slowly do the new findings make their way to hospital clinics, where people are being treated with chemo and radiation—techniques not much less brutal than what Solzhenitsyn described in his novel *Cancer Ward*. The dream teams were trying to cross the divide.

This was part of a broader effort called translational research, which was the subject of a workshop that evening at the Parker House, the grandest of Boston's old hotels. In a room with chandeliers and wainscoted walls, I sat among a group of young scientists who were learning about the different cultures of medical research: biologists who study the chemical cascades inside a cell, clinicians who develop and test new drugs, oncologists and the patients they are treating—they all see cancer in different ways. While the mornings would be filled with lectures, in the afternoons the students would visit cancer clinics and hospital pathology laboratories and watch over a medical ethics panel as it reviewed the rules for conducting new clinical trials—an arena in which the priorities of science and medicine often conflict.

Amy Harmon, a reporter for *The New York Times*, had recently told the story of two cousins with advanced metastatic melanoma, which is about as deadly a cancer as you can get. Both young men—they were in their early twenties—were accepted into a trial for a targeted therapy, vemurafenib, which promised to shrink tumors that were driven by a mutation in a gene called *BRAF*. A small Phase I trial and a larger Phase II had shown promising results. Now it was time for Phase III—675 people in twelve countries—the last step before seeking approval from the FDA.

That is where the dilemma arose. The cousins were lucky to be in the trial—only about half of melanoma cases have this particular mutation. But one of them, Thomas McLaughlin, was randomly assigned to the experimental group, which would get the new therapy ("the superpills," he called them) while the other, Brandon Ryan, was in the control group, which would get dacarbazine, the standard and depressingly ineffective chemotherapy. Both men were dismayed by the arbitrariness of the decision. McLaughlin, whose melanoma was already Stage 4, wanted to switch places with Ryan, whose somewhat less advanced malignancy might have given him a better chance. But that was not allowed. It would compromise the objectivity of the trial.

It was a heart-wrenching story with the good of the few bowing before the good of the many. Without rigorous comparisons like these there might be no new drugs for anyone. Still, it was hard not to think of the people in the control group as sacrificial lambs. Medical ethicists use the term "clinical equipoise" to describe a trial in which there is no a priori reason to consider one treatment superior to another. Only then, many argue, is it right to decide blindly which patient will get which drug. By the time Phase II had ended, vemurafenib appeared to blow dacarbazine out of the water, yet half the patients would now be getting what already seemed like an inferior treatment.

In the end, Phase III proved so definitive that it was interrupted early so both groups could benefit. Initial reports showed that vemurafenib increased progression-free survival, holding the cancer in abeyance for 5.3 months, compared with 1.6 months for dacarbazine. That was enough for the FDA. Before long the drug was approved and being marketed by Genentech. At last report patients were typically living four months longer than those on dacarbazine.

There was no happy ending. Ryan, the cousin in the control group, was among the many who had died during the first year of the trial—sixty-six in the dacarbazine group and forty-two among those getting vemurafenib. By the time another year had passed half of the people who had enrolled in the study were dead. McLaughlin's tumors had spread throughout his body, from his thighs to his brain. But he was still alive and taking the superpills. He told me he was back at his job as a welder, working in the sun. I thought of a passage in *Cancer Ward:* "All the time he was running a race against the tumor to come, but racing in the dark, since he couldn't see where the enemy was. But the enemy was all-seeing, and at the best moment of his life it pounced on him with its fangs. It wasn't a disease, it was a snake. Even its name was snakelike—melanoblastoma." That is an older name for McLaughlin's cancer.

For advanced metastatic melanoma there is nothing that resembles a cure. No matter what the treatment, the aberrant cells discover, through a fortuitous mutation, how to continue with their expansion. Vemurafenib also has a paradoxical side effect: encouraging the growth of other skin cancers, squamous cell carcinoma and keratoacanthoma. Researchers are experimenting with combinations of targeted therapies that aim to overcome these obstacles, hoping that the cancer cells won't develop yet another workaround.

One of the aims of translational research is to bring scientists out of the laboratory so they can see firsthand what patients are going through. At the Parker House, Tom Curran, a professor of pathology at the University of Pennsylvania medical school, described the jarring effect of moving from the isolation of a pharmaceutical company laboratory to St. Jude Children's Research Hospital in Memphis, where he took a job in 1995. Curran had discovered a gene called *reelin,* which helps direct the migration of neurons during the early development of the brain, including the cerebellum. The cerebellum is the center of muscular control and balance, and mice born with defects in the gene walk with a reeling gait. Mutations in developmental genes are also responsible later on for many pediatric cancers, and Curran was particularly interested in medulloblastoma, an aggressive cancer of the cerebellum. Compared with other cancers it is extremely rare: the prevalence among adults is 8 cases in 10 million. But there are 5 cases per 100,000 among children and teenagers, making it the most common pediatric brain tumor. The median age of diagnosis is five. What might begin as nothing more alarming than flulike symptoms can give way to headaches and vomiting, dizziness, loss of balance, and what has been described as "a clumsy, staggered walking pattern."

For Curran, medulloblastoma had been mostly an abstraction until he met children who were being treated for the disease. He knew that for most patients the prognosis was relatively good—the five-year survival rate was as high as 80 percent. For some patients, however, the cancer is recurrent and fatal. Even when the treatments are successful, the side effects can be ruinous. Surgery is usually followed by radiation beamed into the vulnerable brains of children.

"I met one kid, a teenager, who was more than five years free of his disease," Curran told the audience. "He was about sixteen. He was blond haired, blue eyed. He was joking around with the physician. But he was beginning to realize that the rest of his class was continuing to advance and he was leveling off. He began to see that the rest of life was going to be a terrible struggle for him and his family. Working in the lab doesn't give you that kind of perspective. I couldn't get the images out of my head."

He began searching for a better treatment, a drug that would strike at the heart of the cancer without such debilitating effects. First he went to the head of the pathology laboratory at St. Jude and asked for access to the tissue bank, the repository of tumors that had been removed from children over the years. There were only five brain tumors of any kind. He would have to collect his own. Five years passed before he had enough to begin his experiments. By that time research from other labs had emerged suggesting that some medulloblastomas—about 20 percent of all cases—were driven by a genetic defect involving the sonic hedgehog gene. Curran knew the story of the cyclopean lambs, whose birth defects arose from eating lilies with a natural substance—cyclopamine—that repressed the hedgehog pathway. Some cancers, conversely, like basal cell carcinoma and medulloblastoma, appeared to be caused by too much sonic hedgehog activity. Cyclopamine might, in theory, correct the problem and cause the tumors to shrink.

Since cyclopamine was toxic, expensive, and difficult to work with, Curran wanted an alternative. Sitting in a bar after a meeting on brain genetics and development in Taos, New Mexico, he discussed the problem with an authority on hedgehog signaling. He told Curran about some new compounds that were being developed by a biotech company in Massachusetts for the specific purpose of blocking the hedgehog pathway by interfering with a protein called smoothened. Curran went on to show that the substance shrank medulloblastoma tumors in mice. But in younger rodents it also inhibited bone development. Whether the same would occur in children was an open question, but for those with the recurrent form of the disease, facing an early death, the risk might be worth it. Twelve of them enrolled in a clinical trial, and by the time of the Boston workshop there were signs that the drug, vismodegib, was safe and that it was repressing the tumors. Phase II trials were just beginning, but it would still be years before the treatment might be ready for consideration by the FDA. (It has recently been approved for basal cell carcinoma and is also being tested against a few other cancers.)

Vismodegib for medulloblastoma, vemurafenib for melanoma. The names, so weirdly similar, sound like they were spit out by a machine shuffling Scrabble tiles. But they are not without meaning. The suffix "-degib" indicates a hedgehog signaling inhibitor, "vi" comes from "vision" (the drug is "forward looking," a spokesman for Genentech told me), and "smo" from the smoothened protein. As for vemurafenib, "vemu" is for the *BRAF* V600E mutation and "rafenib" means a *raf* gene inhibitor. But the prefixes and often the infixes (the syllables in the middle of the word) are often arbitrary concoctions. Pharmaceutical companies propose the names to a body called the United States Adopted Names Council, which makes the final decision. A researcher told me that the companies pick unwieldy generic names like vemurafenib so that doctors will more readily adopt the catchier brand names, in this case Zelboraf. Vismodegib is sold as Erivedge.

José Baselga, a scientist at Massachusetts General Hospital, described the latest findings on trastuzumab, better known as Herceptin, the drug that seeks out and blocks the HER2 receptor, shutting down signals that encourage cancerous growth. (The suffix "-mab" indicates that it is a monoclonal antibody—a molecule designed to home in on a specific target.) What was now being called "super Herceptin" or trastuzumab emtansine (T-DM1 for short), went a step further, carrying along a cytotoxin and injecting it directly into the malignant cell—chemo delivered right where you want it a molecule at a time. The poison itself is dangerously toxic to the body. But when aimed so precisely, it promised to work as a heat-seeking missile against HER2-positive cancer cells. Now this sounded close to a miracle drug—potent chemo without so many side effects. Baselga said that Herceptin alone had dramatically increased the survival rate for early stage HER2-positive breast cancer from 30 percent ten years earlier to 87.5 percent today. He predicted that super Herceptin, in combination with another

HER2-targeting drug called pertuzumab, could bring this to over 92 percent.

For metastatic cancer the numbers would be much lower, but here too people were hoping for a miracle. Two and a half years after the Boston meeting pertuzumab became Perjeta—another product from Genentech. Combined with Herceptin and old-fashioned chemo it increased progression-free survival—the time before the tumors came back or the patient died—by about six months. As for super Herceptin the wait continued. Some positive results from a clinical trial were used to push for accelerated approval from the FDA, and some patients were outraged when the agency insisted on waiting for Phase III. At a rally outside Boston City Hall, a woman who had been diagnosed with stage 4 HER2 breast cancer five years earlier addressed a small group of people—several of them wore pink T-shirts—and demanded an investigation. "People with the illness need to be in on the discussion. Not just people sitting up in an ivory tower making decisions." They were probably expecting too much. When Phase III results were in, the best that could be said for metastatic breast cancer was that super Herceptin "reduced the risk of cancer worsening or death by 35 percent." The drug did finally receive approval. But for the most aggressive cancers, the state of the art is still measured in months added to what is left of a truncated life.

While waiting for the banquet that followed Baselga's talk, I spoke to a researcher from a southern university about the fact that it was National Breast Cancer Awareness Month and all the attention that was being captured and the money that was being raised. She said she could easily understand why of all malignancies, breast cancer strikes such a deep emotional chord. It is not just because it is one of the most common. Breast cancer is an assault on femininity, female sexuality, and deepest of all, motherhood. But she also seemed a little envious. An unintended consequence of the pink ribbon enthusiasm is to distract from rarer cancers. Her area of research was pancreatic cancer, which has a discouragingly low survival rate. There are often no symptoms. "You go to the doctor with indigestion and find out you have three months to live." Another example of neglected cancers would be UPSC, the one Nancy had.

A whole culture has sprung up around breast cancer. The writer Barbara Ehrenreich, a victim herself, has called it a cult and believes that it trivializes the condition—as though breast cancer were just another life passage, something to get through like menopause or divorce. Not only can you dress in pink, she writes, you can accessorize—with pink rhinestone jewelry. You are told that chemotherapy "smoothes and tightens the skin, helps you lose weight," that baldness is something to celebrate, that your new crop of hair "will be fuller, softer, easier to control, and perhaps a surprising new color."

> As for that lost breast: After reconstruction, why not bring the other one up to speed? Of the more than 50,000 mastectomy patients who opt for reconstruction each year, 17 percent go on, often at the urging of their plastic surgeons, to get additional surgery so that the remaining breast will "match" the more erect and perhaps larger new structure on the other side.

At first Ehrenreich's criticisms sounded harsh to me. Beyond providing comfort and raising money for research, the hope is to get more women into the clinic for yearly mammograms. But it is no longer clear how many lives that saves. More in situ cancers are diagnosed—the tiny, slow growing, "stage zero" tumors a woman is likely to outlive without treatment. But the deadliest cancers can appear so suddenly—within days perhaps after a woman's annual mammogram—and can expand so relentlessly that they often elude detection before spiraling beyond control. A recent epidemiological study of 600,000 women concluded that it is "not clear whether screening does more good than harm." For every life that is prolonged ten women will be treated unnecessarily. But there is no way to know in advance which women those will be.

The same dilemma is faced by men over their own most common cancer, that of the prostate. PSA blood tests can provide early warning, but again they lead to a disturbing number of unnecessary biopsies and surgeries. As with the in situ breast carcinomas, prostate cancers can also smolder harmlessly for decades. About 70 percent of men in their seventies who die of other causes have been found on autopsy to have prostate cancers that they probably knew nothing about. A man rendered impotent and incontinent by surgery may be left to wonder if he should have resisted the pressure to get tested. As with breast cancer, the hype—often well meaning but also driven by the profit motive—has been criticized for overselling the value of early diagnosis. Sports stadiums have become a popular recruiting ground. Urologists offer free tickets in return for an office visit and advertise on arena billboards. A Florida doctor places advertisements on the deoderizer pucks that sit inside the urinals of public restrooms: "Strike Out at the Urinal?" Prostate surgery will increase the flow, though possibly more than you want.

After the meeting at the Parker House I returned to my hotel—a Howard Johnson that was practically adjacent to Fenway Park. The walls and carpets exuded the nicotine smell absorbed from generations of Red Sox fans. I wondered how many of them were going from the stadium with its third-hand smoke to the urologist for a prostate test. I had picked the location because it was near Dana-Farber, where I had an appointment to interview Franziska Michor, who had recently been chosen as one of "the best and brightest" by *Esquire* magazine. She was described as "the Isaac Newton of biology." Michor had a PhD from Harvard in evolutionary biology and her thesis was titled "Evolutionary Dynamics of Cancer." After all the talks about translational science, what I expected to learn about now was the most theoretical of research, vital to understanding the phenomenon of cancer but many stories removed from the clinic.

Random variation and natural selection are the driving forces of cancer as they are of life, and Michor was studying the process with mathematical models. We think of Mendel's peas and Darwin's finches, but it was the quantitative science of population genetics that put our current ideas about evolution—the modern evolutionary synthesis—on sure footing. It was one thing to believe that evolution occurred—that much was soon evident—but could an accumulation of tiny, discrete mutations truly give rise to new species and to the seemingly smooth, gradual changes of evolution? The population geneticists showed with their equations that this was possible, and by the 1930s the modern synthesis was in place. By applying a statistical approach to cancer, researchers in the 1950s uncovered some of the early clues that tumors, like the creatures of the earth, also develop through an accumulation of mutations.

As she sat in her office, Michor described how evolutionary biology and mathematics are making sense of some of the idiosyncrasies of cancer. The revolution in genetic sequencing is making it possible to read out the long list of changes that occur in a cancer cell and even upload them to the Internet. Scientists have been staggered by the numbers, which can reach into the thousands. Most of these, however, are likely to be "hitchhiker" or "passenger" mutations. A cancer cell is one that is mutating wildly beyond control. Many mutations will contribute nothing to the development of the tumor but are simply carried along for the ride. The challenge is to sift through them all and identify the driver mutations, and Michor's lab was working on a model of cancer evolution that she hopes will help make this possible. She was also studying tumors at various stages of their development and trying to figure out the order in which the mutations occur. Is an oncogene mutated first and then a tumor suppressor, or is it the other way around? Perhaps both steps are preceded by damage to a gene essential to DNA repair. Or perhaps there is not one trajectory a cancer cell can follow but many different ones. Knowing a tumor's history might lead to more effective treatments. If a certain mutation tends to arise early on, it would be the one to target. For all its abstract appeal, Michor's work was very much in the spirit of translational research, with the fate of patients not far from her mind.

In another recent paper she and some colleagues considered how oncologists might draw on evolutionary biology to understand how cancer cells can so quickly overcome the obstacles thrown in their path. According to a notion called punctuated equilibrium, championed by paleontologists Niles Eldredge and Stephen Jay Gould, life doesn't always evolve at a steady pace. After long periods of quiet, there can be bursts of genetic innovation. Is that what drives a cancer when, after laying low for a while, it abruptly metastasizes into fresh territory or develops the power to resist the latest chemotherapy?

Ideas from mathematics and evolutionary biology are also being used to show how cancer might be understood through game theory—which was originally devised to find optimal strategies for war. Among the lessons to emerge is that on the battlefield and in the biosphere it sometimes pays for adversaries to cooperate. Robert Axelrod, a political scientist, has suggested how that might apply to competing cancer cells. The evolution of a tumor appears to be a winner-take-all situation. As the cells divide and mutate, one lineage gains the upper hand, developing the hallmarks of cancer, while the others drop out along the way. That seems like a very inefficient battle plan, and Axelrod has proposed an alternative: Some of the cancer cells may evolve the ability to collaborate. Picture two cells sitting side by side. Through a serendipitous mutation the first cell can produce a powerful substance that stimulates its own growth. The other cell lacks this ability, but because of its proximity it is also exposed. It too continues to thrive. As it does, it may learn to synthesize a different product that the first cell lacks. Both will now continue to flourish—at least for a while. Ultimately one lineage may come out on top, but meanwhile the tumor can expand at a rate that wouldn't otherwise be possible.

Not long after my trip to Boston, I sat in on a presentation in which Stand Up to Cancer described its vision of translational research and introduced some of its dream teams. The lecture room was packed and latecomers were turned away at the door. I found a place to stand in back and watched a slickly produced video in which a young woman doing cancer research at the University of North Carolina offered this slogan: "Cancer isn't getting smarter but we are." At first that sounded wrong to me. Inside the body the cancer cells—competing, perhaps cooperating—are continually developing new talents. They evolve the ability to induce angiogenesis and to resist apoptosis and the immune system—and everything else the body throws at them. And once treatment begins, they learn to circumvent the smartest drugs people can devise. No wonder the improvement in survival rates has been so slow. But there is a limit to a cancer's education. Ultimately either the cancer or the patient will die. Either way the evolutionary trajectory is halted. The next cancer must start from scratch.

But what if a cancer could break free? I thought of a recent issue of *Harper's Magazine*. Prominent on the cover were the words "Contagious Cancer" and a painting of a chimerical beast—part bird, part horse, part reptile, part human—dancing in a frenzy with a look of murder on its snaggle-toothed face. It was a painting by the surrealist Max Ernst. It illustrated an article by David Quammen, one of today's best nature writers, and he focused on an affliction discovered in the mid-1990s on the island of Tasmania called devil facial tumor disease. It soon became clear that the lumps—each "an ugly mass, rounded and bulging, like a huge boil"—were being transmitted from one Tasmanian devil to another. This was not through viral infection. When the vicious creatures bit one another's faces, tumor cells were passed along. This was a cancer that had evolved to where it could metastasize to another host. Through genomic sequencing scientists have since traced the origin of the cancer to a single female—"the immortal devil"—whose mutated DNA can be found in all of the tumors.

Another contagious cancer in the animal kingdom is canine transmissible venereal tumor. Again this is not spread by infection but by the direct exchange of cancer cells. In hamsters a different sarcoma can be transferred by injection from one animal to another until the evolving tumor learns to make the jump on its own. It can also be spread between hamsters by mosquitoes.

Quammen described three cases in humans—all medical professionals—in which cancer cells from a laboratory or hospital had become implanted in a wound. In one case, a young woman who poked herself with a syringe acquired colon cancer in her hand. A medical student died of metastatic cancer that began when he pricked himself after withdrawing liquid from a breast cancer patient. Those metastases ended with the recipient. But it is not impossible that a cancer might arise in the wild that has stumbled down an evolutionary pathway that ultimately allows it to leap from person to person. For a cancer like that, its education wouldn't end. It would continue to evolve as it spread across the land. Increment by increment, it would get smarter.

## Chapter 13

# Beware the Echthroi

On a clear winter day I drove the winding road up to the crest of Sandia Mountain, which looms 10,678 feet over Albuquerque, to spend time basking in the emanations of the Steel Forest, a thick stand of blinking broadcast and microwave antennas that serves as a communications hub for New Mexico and the Southwest. Microwaves are a weak form of electromagnetic radiation that sits in the lower half of the spectrum just above radio broadcast waves and below heat waves and the colors of light. Because of the compact size of the waves—half an inch to a foot across—they are easily focused into beams by dish antennas and used to relay television broadcasts, long-distance telephone calls, and other information from tower to tower and to satellites orbiting in the sky.

Microwaves are also transmitted and received by cellular telephones and wireless Internet equipment, and Santa Fe had recently become a nexus for people who believe these emissions cause brain tumors and other sickness. They testified at hearings trying to keep wireless out of the public library and city hall. They opposed every new permit for a cell phone mast—even small ones in church steeples that no one could see. They would know they were there because of their emanations. Or so they believed. One Santa Fean sued his next-door neighbor for remotely poisoning him with her iPhone, and a Los Alamos physicist is sometimes seen in public wearing a chain mail hood for protection. Knowing I was skeptical that the small doses of microwaves the public receives could possibly be harmful, he laid down a challenge. Go to the mountain, he said, and spend an hour or two by the antennas. "See if aspirin cures the headache you'll probably get, and see if you can sleep that night without medication."

After I reached the top, I walked around and admired the endless views, browsed through the gift shop, watched a small outdoor wedding ceremony. I sat for long stretches and read a book about mass hysteria and health scares. The cell phone fears seemed like a prime example, a case of metastatic memes—hard, impenetrable kernels of folk science passed from mind to mind with little deliberation. All the while I had in hand a microwave meter I'd purchased to make sure I was getting a dosage of at least 1 milliwatt per square centimeter. That is the threshold set by the Federal Communications Commission for what it considers safe exposure over a thirty-minute interval. (The sun shines upon us at about 100 milliwatts per square centimeter.) Anti-wireless advocates consider the FCC standard far too high, many times greater than what the brain can bear. After two hours, I drove home and woke up the next morning feeling fine. It might be decades of course before I would know if I had seeded a brain tumor.

If so it would be through a means unknown to science. It is only when you reach the top of the spectrum—the highest frequencies of ultraviolet light, followed by x-rays and gamma rays—that radiation is proven to be carcinogenic. The higher the frequency, the higher the energy—and the smaller and more cutting the waves. Measured in billionths and trillionths of a meter, these are the rays that can zip through cells, tearing electrons from atoms and damaging DNA. Blunter radiation like microwaves can only cause harm by vibrating and heating tissue—that is how a microwave oven boils water and cooks meat. But cell phone and wireless Internet emissions are too feeble for even that. If they were causing cancer, it would have to be in more subtle ways. Electromagnetic fields, microwaves included, can influence the motion of charged particles. And in a living organism streams of charged ions—calcium, potassium, sodium, chloride—flow in and out of cells. So maybe rippling these currents at a particular rhythm somehow elicits malignant behavior, interfering with a crucial cellular pathway by amplifying or squelching it. The oscillations might conceivably suppress the immune system or have epigenetic influences—activating methylation or some other chemical reaction that can affect the output of genes without directly mutating the DNA.

But all of that is speculation. There is no end to laboratory research investigating how the waves might affect mitosis, the expression of DNA, and other cellular functions or alter the efficiency of the blood-brain barrier or enhance known carcinogens. The results are contradictory and inconclusive. One study showed that glucose metabolism, the normal process by which cells turn sugar into energy, was higher in parts of the brain near where a cell phone antenna was held. Whatever clinical significance that might have is unknown, and the study was quickly contradicted by one finding that glucose activity was suppressed. A few studies—the outliers—have hinted that chronic microwave exposure might raise the risk of tumors in laboratory animals. But the experiments are far outnumbered by those finding no effect.

A review by the World Health Organization of approximately 25,000 papers uncovered no convincing evidence that microwaves cause cancer. This is reflected in the epidemiology. For the last twenty years, while cell phone use has steadily increased, the annual age-adjusted incidence of malignant brain tumors has remained extremely low—6.1 cases per 100,000 people, or 0.006 percent—and for the last decade has been slightly but steadily decreasing. That has not kept epidemiologists from investigating whether cell phones might still be having a tiny impact. The most ambitious of these efforts, Interphone, gathered information from five thousand brain tumor patients in thirteen countries and compared it with a control group. No relationship was found between the amount of time talking on a cell phone and the incidence of gliomas, meningiomas, and acoustic neuromas—tumors that occur in areas of the head likely to get the most exposure from mobile phones. There actually was a slightly negative correlation: Regular users appeared to have a lower risk of getting brain tumors than people who didn't use cell phones at all. Rejecting the likelihood of a protective effect, the authors of the final report interpreted the result as a fluke caused by unreliable data, sampling bias, random error—some flaw in methodology. The counterintuitive result also suggested that if there is some effect it is so minuscule that it is swamped by statistical noise.

Interphone was a retrospective study, relying on memory, like the research that had led scientists to believe for a while that eating fruits and vegetables could drastically reduce the incidence of cancer. There was another reason, however, that has kept the results from being accepted as the final word. The study found no sign of a dose-response relationship, where cancer risk would rise steadily according to the number of hours spent on the phone. But for the 10 percent of people who reported the very highest use, the increased risk of glioma appeared to jump abruptly from 0 to 40 percent. A person's odds of being diagnosed with the cancer, the most common of all brain tumors, is about 0.0057 percent. A 40 percent increase would make that 0.008 percent. There was a similar but smaller spike for the other tumors. These were also interpreted by the authors as the result of a methodological flaw. Some subjects reported outlandishly long times spent talking on cell phones—as much as twelve hours a day—and that may have skewed the results.

Maybe people with brain cancer, desperate for an explanation, were overestimating the severity of their cell phone habit. Maybe their memory or their reason was impaired by the tumor. In any case, a later study by the National Cancer Institute looked at gliomas and found no sign that they have been increasing as cell phones have become a ubiquitous part of life. Many epidemiologists were surprised when the International Agency for Research on Cancer decided that there was still enough uncertainty to add microwaves to the long list of possible carcinogens—nowhere near being proven but worth keeping an eye on.

More answers might come from a prospective study that is almost as ambitious as the EPIC project on nutrition and cancer. COSMOS (the Cohort Study of Mobile Phone Use and Health) is monitoring 250,000 volunteer cell phone users for twenty to thirty years, which surely is time enough to find delayed effects. But even when it is completed decades later not everyone will consider the matter resolved. It still can't be said flat out that electrical power lines don't slightly increase the risk of childhood leukemia—a hypothesis that was suggested to widespread disbelief more than three decades ago. The emanations from power lines are many times weaker than even microwaves. Their wavelength is enormous. While the microwaves people have been worrying about are measured in inches and radio broadcast waves in feet—hundreds of feet for the lowest-frequency AM stations—60 hertz power line waves are more than 3,000 miles wide. As they gently roll through neighborhoods they can induce faint currents in whatever they cross, including human cells. No means have been discovered for how that might cause cancer. Over the years most epidemiological studies have turned up no evidence of a danger. But there are always a few anomalies suggesting otherwise.

Sometimes it feels like we're chasing our tails, obsessed with finding causes where there may be none. Robert Weinberg once estimated that every second 4 million of the cells in our body are dividing, copying their DNA. With every division there are imperfections. That is the nature of living in a universe dominated by entropy— the natural tendency for order to give way to disorder. If we lived long enough, Weinberg observes, we all would eventually get cancer. That doesn't mean that we can't reduce the odds, even if only modestly, that we will get cancer before something else kills us. But genetic errors are inevitable and necessary for us to evolve. Evolution is by random variation and selection, and mutations are the grist for the mill. Along the way cells have evolved the ability to identify and repair broken DNA, but if the mechanism was foolproof evolution would stop. The trade-off is probably a threshold amount of cancer.

Robert Austin, a biophysicist at Princeton University, goes so far as to argue that cancer is here "on purpose"—that it is a natural response by which organisms deal with stress. When bacteria are deprived of nutrients, they start replicating and mutating like mad—as though trying to evolve new survival skills. If the source of stress is an antibiotic, the winning adaptation might be one that produces an antidote to the poison—or quickens the pace at which the bacteria can flee. Maybe, Austin proposes, the cells in an organism do the same thing. Backed into a corner they try to mutate their way out of trouble, even if it endangers the rest of the body. The best response might not be to fight back with chemotherapy and radiation, increasing the stress, but to somehow maintain the exuberant cells—the tumor—in a quiescent state, something that can be lived with.

Austin is one of dozens of scientists who have received money from the National Cancer Institute as part of an attempt to break the stalemate in the War on Cancer by importing ideas from outside the usual channels. Franziska Michor, the evolutionary biologist I met in Boston, is also part of the endeavor. In other laboratories, physicists and engineers are bringing their own perspective by studying the mechanical forces involved when cancer cells grow and divide and then migrate through the blood. Instead of speaking the language of biochemistry they use terms like "elasticity," "translational and angular velocity," "shear stress"—as if describing boats leaving dock to navigate down a river. Mathematicians are looking at cells at a different level of abstraction—as communications devices—and using the same ideas from information theory that might be applied in the analysis of radio signals or telephone transmission lines. Perhaps cells can be thought of as oscillators like tuning forks. Malignant ones might be identified by their discordant harmonics—their own special ring. If so there might be a way to retune them. A chemist at Rice University is trying to use radio frequency waves to kill cancer cells. First the cells would be injected with gold or carbon nanoparticles. Then the radio waves would cause them to vibrate, producing enough heat to destroy the cell from inside.

The projects are done in collaboration with oncologists, and a lot of laboratory benchwork is involved. But there are also attempts to step back further and propose whole new theories of cancer. Cell biology is a science of details. There is a grand overarching framework—the modern theory of evolution—but you excel by digging down and mastering thick layers of knowledge about thousands of biochemical gears and the countless ways they can mesh or jam. There are models for how a neuron fires or how DNA is translated into protein. But the closer you look, the more elaborate these mechanisms appear. They are the outgrowth of a long chain of evolutionary accidents, a history that might have spun a different way.

Theoretical physics rewards those who simplify—glossing over details and exceptions and explaining everything in terms of a few big ideas. The lumpers instead of the splitters. The last time I saw Paul Davies, a theoretical physicist and cosmologist, he was speculating on extraterrestrial biology. More recently he and an astrobiologist, Charles Lineweaver, have been playing with the notion that the human genome carries inside its coils an "ancient genetic toolkit"—long buried routines that primitive cells used to form colonies—early precursors to multicellular life. "If you travelled in a time machine back 1 billion years, you would see many clumps of cells resembling modern cancer tumours," Davies ventured. As they join forces to become a malignancy, cancer cells are reenacting this legacy software, "marching to the beat of an ancient drum, recapitulating a billion-year-old lifestyle." When earlier traits long dormant in the genome—hen's teeth, three-toed horse hoofs, vestigial tails in humans—reemerge in later generations, biologists call them atavisms. Cancer, Davies speculates, is an atavistic phenomenon. Stretching out in another direction, he has suggested that the transition of a healthy cell to a cancerous one may have something to do with quantum physics.

It was surprising to see Davies brainstorming about cancer. Even more unexpected was Daniel Hillis, a computer scientist and roboticist who is heading a team at the University of Southern California that is assembling detailed computer simulations of cancer—virtual tumors—that might be used to predict which drugs work best. I'd first heard of Hillis when as a student at MIT he helped build a Tinkertoy computer that played tic-tac-toe. He went on to start a company called Thinking Machines. He may be best known as the designer of a giant clock that is being assembled inside a mountain in West Texas, where it is supposed to keep running for ten thousand years, chiming through the millennia even if the human race is gone. At a session organized by the NCI he told an audience of oncologists that the way they were fighting cancer is all wrong—that we need to think of cancer as a process, not a thing. A body does not have cancer, it is "cancering." Treatment should focus not on attacking a specific type of tumor in a specific organ but on looking at the patient as a complex system. Somewhere in the network of interlocking parts—the immune system, the endocrine system, the nervous system, the circulatory system—something has become unbalanced, and for every patient there may be a different way to set it right. That might have struck some listeners as so much holistic fuzz. But Hillis has been pursuing the idea by building another of his ambitious machines. Instead of the genome he was concentrating on the proteome—all of the proteins that are present in a cell at any one moment. Reading the genome gives you the instructions for making each of the cell's working parts. Reading the proteome shows which parts are actually being made and in what abundance—a snapshot of the state of the system.

Scientists have been working for years on mapping the proteome—a formidable task involving laboratory techniques like liquid chromatography and mass spectrometry. In collaboration with

David Agus, an oncologist, Hillis started a company that is trying to automate the multiple steps with a robotic assembly line. Given a drop of blood, the machine extracts and sorts the proteins, arranging them in an image that looks like stars in a sky. Each kind of protein appears as an illuminated spot, and its brightness shows how much there is.

Suppose you have two patients with the same kind of cancer. One responds to a drug and the other does not. Using a device like Hillis's, you could take their proteomic snapshots and lay one on top of the other and look for something that is different. Even if you don't know what the pattern means, it might be used as a marker to identify which patients will most likely benefit from the drug. I was reminded of Henrietta Leavitt, the astronomer who had died of stomach cancer but not before discovering Cepheid variables, the pulsating stars cosmologists use to measure the universe. She would start with two images of the same patch of sky—glass photographic plates taken a few weeks apart. One would be a negative with the stars glowing in black. She would place that plate on top of the other and hold the glass sandwich to the light. Stars that had grown brighter would appear as larger white spots with smaller black centers. On a plate taken weeks later the white spot would have shrunk to its previous size. No one yet knew the physics that caused the stars to blink, but she was able to correlate their rhythm with their distance from the earth. Sometimes our eyes can glimpse connections that our brains don't understand.

As the population ages, cancer is outrunning us. But placed under this stress we are like those madly replicating bacteria Austin talked about—spinning out combinations of memes instead of genes. New ideas. Maybe we really are getting smarter than cancer. Efforts like the Cancer Genome Atlas are continually announcing new discoveries—zeroing in on the genetic details of cancers and sorting them into subtypes, each one potentially vulnerable to a different treatment. As the information multiples, custom therapies will be further customized. Targeted drugs will become ever more precise.

When a tumor finds a workaround, other drugs will be ready to go after the new mutation. Pursuing a different strategy, a new class of pharmaceuticals will switch back on apoptosis. Immune system boosters will learn to cleanly distinguish between what is a tumor and what is healthy flesh. A cocktail of these advanced treatments will stop cancer—even advanced metastatic cancer—in its tracks or manage it indefinitely as a chronic disease. Or maybe in ten years we will be reading how these approaches too are falling behind in the cellular arms race and we will be forced to look at cancer in an entirely different way.

About a year after he showed me his lab in Princeton, Austin was invited to Davies's domain at Arizona State University to give a talk called "Ten Crazy Ideas About Cancer." In the end he came up with five, and one in particular has stuck in my mind. It was about mitochondria. I remembered my surprise when I learned years ago that the mitochondria, these tiny things inside our cells, might once have been bacteria—individual creatures that became trapped somehow. The mitochondria have their own DNA and can replicate independently within the cytoplasm. With their ability to burn glucose and power the Krebs cycle—the chemical dynamo that energizes the cell—these symbionts provided their hosts with an evolutionary advantage. They have also long been suspected of playing a part in cancer. Mutations to the mitochondrial DNA are found in many different tumors. That might just be collateral damage from the havoc of a cell careening toward malignancy. But there are reasons to think the mitochondria are more directly involved. For one thing they help initiate apoptosis, the cellular suicide routine. In his crazy ideas talk, Austin speculated that cancer might begin when the mitochondrial symbionts rebel. From the wear and tear of generating energy, they become damaged and spew out free radicals that eat at other parts of the cell, including the genome. The cell gets sicker and the only recourse is to destroy itself. But the mitochondria refuse to cooperate. They don't want to die. More mutations follow and the cell becomes malignant.

The picture Austin drew reminded me of *A Wind in the Door,* an allegorical novel by Madeleine L'Engle in which forces of good and evil contend over the universe. It is a sequel to *A Wrinkle in Time,* which I discovered as a boy in my junior high school library. It was in L'Engle's fantasy that I first came across the idea of a tesseract— a four-dimensional cube. The idea blew my eighth grade mind. *A Wind in the Door* is even stranger. This time Charles Wallace, the precocious young protagonist, is suffering from a degenerative disease. His mitochondria are gravely wounded, and his micro-biologist mother discovers the cause. There are symbionts within the symbionts—the fictional farandolae—and they are rebelling. They are egged on by the Echthroi, supernatural agents of entropy. Swooping through the universe, they destroy order by what they call Xing—unnaming things, eating information. Charles Wallace and his sister beat back the demons and after a trip inside a mitochon-drion the boy is saved. But in the real world the Echthroi are always with us, stripping off labels, dedifferentiating cells, freeing them to make cancer.

Early in the spring, a year after the Relay for Life and a year after the last of Nancy's radiation, we traveled to Patagonia to celebrate. There was a lodge on a lake in the mountains, and for years it had been high on our list of places to see. We wouldn't be roughing it. Each evening the guests were served fine dinners with good Chilean wines. Our room, through sheer luck, was the best in the house, with a view of both the lake and a waterfall. But the luxury wasn't the main draw. Each morning we would depart with a group on hiking expeditions to glaciers, mountains, lakes, and rivers. Nancy looked so thin and frail to me, but she made it to the very end of every hike.

One evening after dinner we walked out of the lodge and the stars were more brilliant than we had ever seen. Brilliant and strange. The constellations were unfamiliar, and a pair of dwarf galaxies stared down at us like two big eyes. It took a minute to realize that they were the Magellanic Clouds. Magellan had used them to navigate in the Southern Hemisphere, where the North Star is not visible. And it was within these starry nebulae that Leavitt discovered the Cepheids. Had she lived in this century, the statisticians tell us, her odds of getting stomach cancer would have been much lower. But it still probably would have killed her. With few symptoms at first, it is another of those cancers that is often not noticed until it has metastasized. Chemotherapy or radiation can only hold it in abeyance. For all our understanding of cellular science there is still so much progress to be made. But there are occasionally good surprises. Nancy's odds hadn't been good either, and soon she was thriving. Back in Santa Fe she bought a new bicycle and rode in the Santa Fe Century, covering fifty miles.

Every few months she would go to the cancer center for a blood workup. They were keeping their eye on her level of CA-125, a protein that is used as a biomarker for the presence of endometrial and other cancers. Too much CA-125 doesn't necessarily mean the cancer is back, and you can have cancer without elevated CA-125. It's a blunt-edged tool, but in any case Nancy's remained normal. She also had a PET scan twice a year, and every time she was clear.

In the fifth year after cancer she bought a horse—something she had wanted to do since she was a girl—and then another horse, and in the sixth year A.C., as she called it, she had fallen in love with two and a half acres of land on the far side of town. It had barns and stables and bordered a square mile of open land. She was determined not to waste a day of a future she had almost lost. It wasn't an expensive piece of property, and she had inherited a little money after her mother died of breast cancer. So we took out another mortgage and bought the land, and she rode her horses there whenever she could. We called it the ranch.

I wasn't a rider but I became obsessed with combating the weeds. There were hordes of the nastiest varieties. In the gardens at home I confronted an occasional kochia. Here they were everywhere. Even worse was a cousin of the weed—another invader from the Russian steppes—called *Salsola tragus,* or tumbleweed. Perversely embraced as an icon of the Old West, it first came to South Dakota in the late 1800s possibly from the Ukraine. I imagined it arriving as a seed stuck to an immigrant's sock. Then it began spreading everywhere. Some farmers thought it was part of a conspiracy and gave it another name, Russian thistle. On the Nevada test range, after aboveground nuclear explosions were banned, salsola was the first life to come back.

I tried everything but ionizing radiation to eradicate it. Early in the spring the plants began to appear as tiny bluish-green stars. I learned to recognize them immediately, surgically removing them with a hoe. When that task became overwhelming I burned them with a weed torch. And still they would appear and grow larger, developing ugly lizard-like purple-striped stems. The stems would grow into a tangle bristling with thousands of thorny seeds. A single tumbleweed could have a quarter million of them. I bought a book on weed science and picked the best chemo—an herbicide called 3,5,6-trichloro-2-pyridinyloxyacetic acid, or triclopyr. It was said to break down rapidly in the soil so the environmental impact was low, and it was selective, killing various kinds of weeds but not the native grasses we wanted to encourage. Spray it on a plant and it travels through the phloem and concentrates in the rapidly proliferating cells of the meristem. There it is believed to mimic plant growth hormones called auxins. Throwing this wrench into the machinery causes the new stems to grow stunted and gnarled, and the plant soon dies. It looks like it is writhing in agony. It was like chemo in reverse, inducing something like cancer. I was careful as I sprayed, in case the fine print was mistaken when it said triclopyr was not a human mutagen or otherwise known to be carcinogenic. It decomposed rapidly enough that it was not believed to hurt wildlife or to pollute the water table.

For all that work Russian thistle continued to spring up by the dozens. When Nancy wasn't working or caring for her horses and when I wasn't writing, we would walk every square foot of the land pulling up the weeds by their roots. Every weekend we filled big plastic garbage bags with hundreds of them and I hauled them to the dump. The hope was to grab every one before it set seed—to break the longstanding cycle. In the spring dead skeletons of Russian thistle would tumble in from afar, but we hoped to reach an equilibrium, something we could control. It was a relief when winter came and everything stopped growing.

Come spring we anxiously surveyed the land. It seemed clean at first, then patch by patch the evil little stars reappeared and the battle resumed. I began to notice that the seedlings hid from me beneath junipers, crouched almost invisible next to fence posts and rocks. And when I did spot them, just an inch or two high, some of them were already producing seeds—stealthily reproducing before I could stop them. They seemed to be adapting to me, evolving before my eyes.

There is an old thought experiment in physics involving Maxwell's demon, an imaginary little creature who tries to overcome the universe's inevitable march to disorder by catching wandering molecules and nimbly returning them to their proper location. Replacing each fallen grain in a slumping sand castle. Pulling every weed in a pasture. Repairing every mutation in a cell's DNA. With effort entropy can be forestalled—life itself consists of vessels of order swimming against the entropic tide. With our tools and intelligence we can strike small victories and hold off death for a while. But it is the tide that will eventually prevail. Try as he might, Maxwell's demon will ultimately be defeated. In the end, the Echthroi always win.

# Joe's Cancer

> *A life-view by the living can only be provisional.*
> *Perspectives are altered by the fact of being drawn;*
> *description solidifies the past and creates a gravita-*
> *tional body that wasn't there before. A background of*
> *dark matter—all that is not said—remains, buzzing.*
>
> —JOHN UPDIKE, *Self-Consciousness*

The next spring at the ranch, I'm told, the salsola was as bad as ever. I wasn't there to see it. During that year our marriage ended, seventeen years after it had begun. For a long time our lives had been diverging. The cancer had brought us closer but now it was gone. Brushing so near death makes a person think about how she wants to spend what is left of life. Nancy had her reasons for deciding it was not with me.

Around that time I received an e-mail from my youngest brother, Joe. He was on the road between his home in Dallas and Albuquerque, driving one of his daughters back to college. Somewhere in the expanse of eastern New Mexico he was chewing on a snack when suddenly he heard a loud crack and a searing pain ripped through his jaw. He continued on to Albuquerque and lay awake all night before flying home to see his doctor.

Although he hadn't talked much about it, Joe had been having trouble with his mouth for years. It had begun when a white area appeared on the gums of his lower left jaw. A biopsy found abnormal cells that were described as precancerous. Nothing to worry about— just to keep an eye on as you might a suspicious mole. The problem didn't arise again until three years later when he felt a soreness. It was also in his lower left jaw. During the next few months, a dentist, an internist, and an oral surgeon each concluded that the best course was to wait and see. Which is what he did until the pain got worse and he was found to have an abscess where a wisdom tooth had been extracted. There was also bone resorption in the area and a couple of dying teeth. All of this was on the left side of his mouth. The skeletal erosion was shored up with bone grafts, the two teeth were extracted, and work began on placing artificial implants. Meanwhile the pain in his jaw continued and was soon accompanied by a ringing in his ears and a sore throat. An otolaryngologist prescribed an antibiotic mouthwash. He had another bone graft and then, not long after, came the incident on the highway.

The next day in Dallas Joe was given a CT scan and told that he had dislocated his jaw—that all the dental work had caused him to chew in a way that had twisted the bone from its socket. It seemed like a plausible explanation. The doctor prescribed muscle relaxants and, as Joe put it, soft, squishy foods. The white area first noticed three years before was still there and had grown larger. A stabbing pain in his ear led to another CT scan and, for the first time, an MRI. An MRI, I later learned, may be more likely to see abnormalities in soft tissue. And there it was. Inside his mouth and beneath the skin, a tumor—about an inch long—was eating into my brother's jawbone. A biopsy identified it as squamous cell carcinoma, the same cancer Percivall Pott found in the chimney sweeps and that Katsusaburo Yamagiwa induced by applying coal tar to rabbits' ears. A PET scan showed that it hadn't spread anywhere else in his body. Holding on to that fact, Joe sent my siblings and me an e-mail with the subject line: "Good News!" That is the kind of man he was.

There was so much more information about this cancer than there had been for Nancy's carcinoma. Squamous cells form the outer layers of the epidermis, the envelope of flesh that is exposed to the world. Beneath them are the basal cells. As skin cells die and slough off, it is the basal cells that divide and produce replacements. These push upward to form the outer skin. Carcinomas of the basal cells are the ones that are usually harmless. Years ago I'd had one removed from the side of my nose. Carcinoma in the squamous region is more aggressive, but it still has a relatively high survival rate, especially if caught early on. What was occupying Joe's body is referred to specifically as squamous cell cancer of the head and neck, and the National Cancer Institute said that about 52,000 people would be diagnosed with it that year. Why he was one of them was another mystery. Other than being male and over the age of fifty, he had none of the risk factors. He drank alcohol sparingly and he had never smoked. He did not chew betel nuts, which is offered as an explanation for the high rate of the cancer in Southeast Asia. He was tested and found clear of HPV, another possible factor.

Surgery lasted eight hours and was mostly a success. The tumor was now 2.5 inches long—more than twice the size measured a few weeks earlier in the MRI—and was found to be wrapped around a nerve. That accounted for the racking pain. The mass was successfully extracted along with the damaged part of his jaw. While that was happening a piece of bone was removed from his hip to fill the gap. In the end it couldn't be used. That much of the surgery was a waste. The arteries in the transplant site had collapsed so that there wasn't enough blood flow to support a graft. Later on there would have to be another operation. But for now the important thing was that the cancerous tissue was apparently gone. Of thirty-one lymph nodes that had been removed, only one was found to have been affected. Maybe it had done its job and kept the malignant cells from moving farther. Their next stop is usually the lungs.

With a tracheotomy to help him breathe and a feeding tube temporarily threaded through his nose, Joe recovered for nine days in the hospital and then went home. Next would come six weeks of chemo (cisplatin and Erbitux, a monoclonal antibody) and radiation. He would also be given a medication meant to protect his saliva glands from burns, and a feeding tube would be installed in his stomach. He was taking this all with a humbling equanimity even when, just before treatment began, he noticed a swelling. A new tumor was growing, this one in his upper left jaw. Another was found near his Adam's apple.

"I always thought that hearing the words 'you have cancer' were the worst you could hear," he told us, "but I was wrong. 'We found more tumors' is much worse. . . . I think I now realize just what a vile, evil sickness cancer is. The doctors keep chasing it around the body."

I thought again of Solzhenitsyn's *Cancer Ward,* where one of the patients is speaking in awed resignation about his own malignancy: "A melanoblastoma is such a swine you only have to touch it with a knife and it produces secondaries. You see, it wants to live too, in its way." The recurrences of Joe's cancer were happening in spots that had been disturbed by the surgery, and the doctors thought the new tumors might have somehow been seeded during the operation. But there were other possibilities. I found a paper from 1953 describing a concept called field cancerization—multiple primaries springing up in the same location at about the same time. It was possible that malignant cells from the original tumor could spread to nearby locations. But studies suggested that, in cases like Joe's, each tumor might have developed independently. That seemed like an incredible coincidence, but there were ways it might happen. Researchers had discovered that tissue lying between squamous cell tumors— tissue that otherwise appeared normal—had genetic abnormalities, including mutations to the tumor suppressor gene *p53*. The mouth and throat are steadily exposed to carcinogens. A cell damaged by a mutagen would give birth to progeny all with the same defect. One of those might sustain another hit and give rise to a family of doubly mutated cells. In time, as the cells kept dividing, there would be a field of precancerous cells, all with multiple mutations and each awaiting the final push. Another possibility is that the cancer field was created early on during development, when a mutated mother cell gave rise to progeny that were all dispatched to form the lining of the mouth and throat. From the very beginning these cells would share the same abnormalities—a head start in becoming cancerous. However the field came to exist it would be lying there primed—"a ticking time bomb" one paper called it—for multiple cancers. Yet it still seemed strange that so many cells could acquire that final precipitous mutation—all at about the same time—especially in someone who wasn't smoking or drinking or chewing betel nuts.

After the initial shock, Joe accepted the news as another setback. It just meant expanding the target of the radiation and altering the chemo. He believed deeply in God and his doctors, and his wife and daughters kept him looking ahead. "I have a virtual army of people praying for my health and recovery—from all cultures, all denominations," he wrote on a webpage his family maintained to keep people apprised of the news. "I have no doubt I will get rid of this cancer and will be back to normal—whatever it takes to do it. I am so blessed that I know in the depths of my soul that I will defeat this." It would be good to believe in God. After two weeks of treatment his patience was rewarded. "Good news! End of week two and one of the tumors is GONE! Just a hole where it used to be. The others should soon follow."

His treatment was as rough as it gets. Twice he returned to the hospital with nausea and dehydration—the results of an infection. But he passed the halfway point—he told us he saw the light at the end of the tunnel—and then began the recovery from the cure. He felt better than he had in months and was happy how quickly he was able to work again from home. And then he got pneumonia, and at the hospital the doctors noticed a mass near his esophagus. It might just be mucous, they assured him. Of course it was a new tumor. As he prepared for six more weeks of radiation and chemo, another tumor appeared in his jaw. The doctors said that one too could be treated. "Great news!" Joe wrote again. "Longer time in treatment, but it can be beat!"

It had been January when he heard that scary cracking sound while he was eating. By mid-October he had finished the second round of treatment. During that time he exhausted his sick leave. His boss tried to get him another extension, but soon Joe was unemployed. You can fire someone because they have cancer. Joe said he understood. He was sure he would get his job back once he was well. More than a month passed before he felt a new soreness, this time around his collarbone. "Buckle your seatbelt," he said. On Thanksgiving Day he wrote from the hospital:

"I have so much to be thankful for this year. . . . I had surgery yesterday, to help my breathing (by removing dead cells that were left over from radiation) and to biopsy the clavicle growth. Good news on both fronts. I am back to normal breathing! And the tumor close to the clavicle can be treated with radiation! I'm waiting to go home now." There was no such thing to him as completely bad news.

But then came more tumors, too many now to radiate. A body can stand only so much. "We can shrink them as much as we can with straight chemo," Joe reported, "but this will not kill them. I don't know if I have 6 months or 6 years." That was November 30. He didn't even have six weeks.

He spent Christmas at home with his family. The chemo now was doing as much harm as the cancer, so the doctors had stopped all of his medications except for those to control pain. If he regained his strength, they said, treatment could always resume. We tried to believe that might really happen. He was lethargic and having convulsions, but just after Christmas he woke up clearheaded and feeling better than he had for days. He smiled at his wife and took her arm, looked in her eyes and said, "When?" Then he fell asleep. It was like in a movie, she later said. He reawoke and his daughters came into the room. They were all laughing together and he was telling them he loved them. He was Joe again. And before they knew it he was gone.

At his memorial service, the minister talked about the mystery of death, the love that cancer can never take away, the power of God to unbind and set free. He told how Joe had sent him an e-mail the morning before surgery. He said he felt like Commander Adama on the science-fiction series *Battlestar Galactica*. He was going in to remove the invader.

In my collection of old scientific instruments is a device called a spinthariscope, a name that comes from the Greek word for spark. It looks like a brass eyepiece from an old-fashioned microscope, and on the side is engraved "W. Crookes 1903." That is the year William Crookes, the inventor, unveiled it at a gala held by the Royal Society. I doubt Crookes made this one—there are a number of spinthariscopes with the same engraving still floating around on the market. Maybe it was issued as part of some commemorative event. Inside the brass tube a piece of radium is mounted next to a screen of zinc sulfide—the phosphorescent chemical that was mixed to make the glowing paint that poisoned the Radium Girls. As the radium decays it shoots out alpha particles and they are registered as tiny flashes of light. Each flash is from the disintegration of the nucleus of a single radium atom, and you can watch the show through a lens on the other end of the instrument. The effect is mesmerizing. Crookes compared it to "a turbulent, luminous sea." Sometimes when I cannot sleep I pick up the device from the bedstand and watch the random light bursts—these miniature nuclear explosions. I think about the randomness of the mutations that cause cancer, and about the fact that I am holding something radioactive so close to my eye. The alpha particles are safely contained inside the instrument, but if I scraped out a speck of radium and swallowed it, I might die. How can life be so robust and yet so delicate?

The flashes from the decaying atoms are the purest kind of randomness. According to a bedrock law of nature—quantum mechanics—there is no way to predict when a single nucleus will decay. As long and as hard as you look into the spinthariscope you will never discern a pattern. Nor can you find a reason why one particular radium atom shoots out an alpha particle at this particular moment and not the next. Two identical nuclei are sitting side by side, and suddenly one will decay for no reason whatsoever, leaving the second one to sit there for another thousand years. What can only be predicted is how a mass of radium—a population—will behave. Approximately half the nuclei will decay over a span of sixteen centuries. But we can never know which ones.

That is how it is with cancer. Given a large enough group of people, we can predict what percentage of them will be stricken but we cannot know who they will be. This is not irreducible randomness like that inside atoms. With enough information—demographic, geographic, behavioral, dietary—we can narrow the pool of those at risk for certain cancers. In the future genomic and proteomic scans and technologies not yet known may allow the pools to be narrowed further still. But there is only so far we can go. Whether any one person gets cancer or does not will always remain mostly random.

I place the spinthariscope back on the table. There is no way to turn it off. All night and all day the flashes continue, year after year unseen. The radium itself will continue decaying for centuries, but first the scintillation screen and the glass lens will wear out. Maybe the brass will survive for archaeologists to wonder over like ancient coins. I imagine how my yard will look by then if there are no people attending to it. First the weeds will take over, crowding out the less aggressive life. Leaves will blow onto the patio, slowly disintegrating to make soil in which more weeds will grow. Seeds of Siberian elm—the unkillable weed trees that have spread across the West (more Echthroi from Eurasia)—will become wedged into cracks in the concrete, slowly splitting it apart as they grow. The cracks will gradually widen and the roots will creep beneath the foundation of my house and eventually it will fall. I think of paintings in museums of grand Roman ruins covered with vegetation, slowly digested back into the ground.

Inside my body, 10 trillion cells (these tiny Maxwell's demons) are battling the same inevitable slump toward entropy. It is eerie to think that inside each one—invisible to the eye—so much is happening. The cell doesn't *know* it has DNA or RNA or telomeres or mitochondria. It doesn't know that A fits with T and C with G. Or that CTG stands for the amino acid leucine, or GCT for alanine—these molecular beads strung together to make proteins. There are no labels, no genetic alphabet written anywhere. There are no instructions. Somehow it all just works. And when it doesn't we rage against the machine.

# Notes

The unwieldy Internet addresses called URLs (for "uniform resource locater") were never meant to appear in print. They are behind-the-scenes directions for a computer mindlessly executing the click of a mouse on a hyperlink. I have excluded them from the notes section of the printed editions of this book. For almost every paper I have cited, the abstract and often the complete text can be found easily with an Internet search, which is much easier than typing, letter by letter, slash by slash, the precise URL. The webpages I refer to can also be readily located. All were accessible as this book was going to press. An online version of the notes, available through my website, talaya.net, will carry links to all of the references, as will the book's electronic editions.

CHAPTER 1  Jurassic Cancer

5  **the Dinosaur Diamond Prehistoric Highway:** My road trip to Colorado and Utah was in September 2010. For a description of the Morrison Formation and Jurassic Colorado see Ron Blakely and Wayne Ranney, *Ancient Landscapes of the Colorado Plateau* (Grand Canyon, AZ: Grand Canyon Association, 2008); John Foster, *Jurassic West: The Dinosaurs of the Morrison Formation and Their World* (Bloomington: Indiana University Press, 2007); "Reconstructing the Ancient Earth," Colorado Plateau Geosystems website, last modified July 2011; and Ron Blakely, e-mail message to author, March 9, 2012.

5  **Giant termite nests:** Stephen T. Hasiotis, "Reconnaissance of Upper Jurassic Morrison Formation Ichnofossils, Rocky Mountain Region, USA," *Sedimentary Geology* 167, nos. 3–4 (May 15, 2004): 177–268 (reference is on 222–23); and Hasiotis, e-mail message to author, March 9, 2012.

6  **Dinosaur Hill:** The discovery of the *Apatosaurus* skeleton is described on two interpretative signs at the site.

7  **it caught the eye of a doctor:** The story of Raymond Bunge was told to me by Brian Witzke in an e-mail, August 3, 2010. Some details about Bunge's life are in "Papers of Raymond Bunge: Biographical Note," 2011, University of Iowa Libraries Special Collections and University Archives website.

7  **an attractive chunk, 5 inches thick:** Bruce M. Rothschild, Brian J. Witzke, and Israel Hershkovitz, "Metastatic Cancer in the Jurassic," *Lancet* 354 (July 1999): 398. More details were provided in e-mails from Rothschild in June 2010, October 2010, November 2010, and July 2011.

8  **layered, onion-skin look:** Rothschild, Witzke, and Hershkovitz, "Metastatic Cancer."

9  **scattered references . . . to other dinosaur tumors:** Rothschild, Witzke, and Hershkovitz, "Metastatic Cancer."

9  **"This observation extends recognition":** Rothschild, Witzke, and Hershkovitz, "Metastatic Cancer."

10  **tumors that arise from misguided germ cells:** See, for example, Naohiko Kuno et al., "Mature Ovarian Cystic Teratoma with a Highly Differentiated Homunculus: A Case Report," *Birth Defects Research. Part A, Clinical and Molecular Teratology* 70, no. 1 (January 2004): 40–46.

11  **With a portable fluoroscope:** B. M. Rothschild et al., "Epidemiologic Study of Tumors in Dinosaurs," *Die Naturwissenschaften* 90, no. 11 (November 2003): 495–500.

11  **a picture of him wearing a dinosaur T-shirt:** John Whitfield, "Dinosaurs Got Cancer," *Nature News* 21 (October 2003), published online October 21, 2003.

12  **may have been more warm-blooded:** Rothschild's "Epidemiologic Study of Tumors in Dinosaurs" cites the work of Anusuya Chinsamy, including A. Chinsamy and P. Dodson, "Inside a Dinosaur Bone," *American Scientist* 83 (1995): 174–80.

12  *Edmontosaurus* **"mummies":** Phillip L. Manning et al., "Mineralized Soft-Tissue Structure and Chemistry in a Mummified Hadrosaur from the Hell Creek Formation, North Dakota (USA)," *Proceedings of the Royal Society B: Biological Sciences* 276, no. 1672 (October 7, 2009): 3429–37.

13  **Rothschild considered the odds:** L. C. Natarajan, B. M. Rothschild, et al., "Bone Cancer Rates in Dinosaurs Compared with Modern Vertebrates," *Transactions of the Kansas Academy of Science* 110 (2007): 155–58.

13  **Rothschild and his wife:** Bruce M. Rothschild and Christine Rothschild, "Comparison of Radiologic and Gross Examination for Detection of Cancer in Defleshed Skeletons," *American Journal of Physical Anthropology* 96, no. 4 (April 1, 1995): 357–63.

13  **Autopsies at the San Diego Zoo:** M. Effron, L. Griner, and K. Benirschke,

"Nature and Rate of Neoplasia Found in Captive Wild Mammals, Birds, and Reptiles at Necropsy," *Journal of the National Cancer Institute* 59, no. 1 (July 1977): 185–98.

14 **paleontologists in South Dakota:** They were at the Black Hills Institute of Geological Research.

14 **"a weird mass of black material":** John Pickrell, "First Dinosaur Brain Tumor Found, Experts Suggest," *National Geographic News,* November 24, 2003, published online October 28, 2010.

14 **"It certainly would take a bizarre event":** Pickrell, "First Dinosaur Brain Tumor."

15 **300-million-year pile of geology:** That is approximately when the bottommost layers of the plateau, the Morgan Formation and Weber Sandstone, were formed. See Halka Chronic and Lucy M. Chronic, *Pages of Stone: Geology of the Grand Canyon and Plateau Country National Parks and Monuments* (Seattle: Mountaineers Books, 2004), 90. I also referred to Halka Chronic's book *Roadside Geology of Colorado* (Seattle: Mountaineers Books, 2004) and to Annabelle Foos and Joseph Hannibal, "Geology of Dinosaur National Monument," Cleveland Museum of Natural History (1999), published online by the National Park Service. Two pamphlets written and illustrated by Linda West and published by the Dinosaur Nature Association (Jensen, Utah) were also used as sources: *Journey Through Time: A Guide to the Harper's Corner Scenic Drive* (1986) and *Harper's Corner Trail* (1977).

15 **jawbone of a primitive armored fish:** Luigi L. Capasso, "Antiquity of Cancer," *International Journal of Cancer* 113, no. 1 (January 1, 2005): 2–13.

15 **Laramide orogeny:** For a beautiful account, see John McPhee, *Rising from the Plains* (New York: Farrar, Straus and Giroux, 1986), 43–55.

15 **ancient elephants, mammoths, and horses:** Capasso, "Antiquity of Cancer."

15 **Hyperostosis, or runaway bone growth:** Capasso, "Antiquity of Cancer"; and Raúl A Ruggiero and Oscar D Bustuoabad, "The Biological Sense of Cancer: A Hypothesis," *Theoretical Biology & Medical Modelling* 3 (2006): 43.

16 **an ancient buffalo and an ancient ibex:** Capasso, "Antiquity of Cancer."

16 **cancer in the mummy of an ancient Egyptian baboon:** Alexander Haddow, "Historical Notes on Cancer from the MSS. of Louis Westenra Sambon," *Proceedings of the Royal Society of Medicine* 29, no. 9 (July 1936): 1015–28.

16 **Even a lone single-celled bacterium:** Jules J. Berman, *Neoplasms: Principles of Development and Diversity* (Sudbury, MA: Jones and Bartlett Publishers, 2009), 67–69.

16 **A bacterium called *Agrobacterium tumefaciens:*** M. D. Chilton et al., "Stable Incorporation of Plasmid DNA into Higher Plant Cells: The Molecular Basis of Crown Gall Tumorigenesis," *Cell* 11, no. 2 (June 1977): 263–71.

16 **A remarkable paper:** Philip R. White and Armin C. Braun, "A Cancerous Neoplasm of Plants. Autonomous Bacteria-Free Crown-Gall Tissue," *Cancer Research* 2, no. 9 (1942): 597–617.

16 **larval cells can give rise to invasive tumors:** Berman, *Neoplasms,* 69–70.

17 **carp, codfish, skate rays:** Capasso, "Antiquity of Cancer."

17 **Trout, like people, get liver cancer:** Berman, *Neoplasms,* 71.

17 **sharks do get cancer:** Gary K. Ostrander et al., "Shark Cartilage, Cancer and the Growing Threat of Pseudoscience," *Cancer Research* 64, no. 23 (December 1, 2004): 8485–91.

17 **parathyroid adenoma in turtles:** Capasso, "Antiquity of Cancer."

17 **Amphibians are also susceptible:** Berman, *Neoplasms,* 71.

17 **a strange variation on the theme:** Charles Breedis, "Induction of Accessory Limbs and of Sarcoma in the Newt (*Triturus viridescens*) with Carcinogenic Substances," *Cancer Research* 12, no. 12 (December 1, 1952): 861–66.

17 **Could this be another clue:** Richmond T. Prehn, "Regeneration versus Neoplastic Growth," *Carcinogenesis* 18, no. 8 (1997):1439–44.

17 **Mammals appear to get more cancer than reptiles or fish:** See, for example, Effron, Griner, and Benirschke, "Nature and Rate of Neoplasia."

17 **Domesticated animals seem to get more cancer:** Capasso, "Antiquity of Cancer."

18 **a close relationship between size and life span:** See, for example, John R. Speakman, "Body Size, Energy Metabolism and Lifespan," *Journal of Experimental Biology* 208, no. 9 (May 2005): 1717–30. For a deeper look at scaling phenomena see James H. Brown and Geoffrey B. West, *Scaling in Biology,* Santa Fe Institute Studies on the Sciences of Complexity (New York: Oxford University Press, 2000).

18 **Peto's paradox:** R. Peto et al., "Cancer and Aging in Mice and Men," *British Journal of Cancer* 32, no. 4 (October 1975): 411–26.

18 **The mystery was succinctly posed:** John D. Nagy, Erin M. Victor, and Jenese H. Cropper, "Why Don't All Whales Have Cancer? A Novel Hypothesis Resolving Peto's Paradox," *Integrative and Comparative Biology* 47, no. 2 (2007): 317–28.

18 **roughly a billion heartbeats:** I first wrote about this in "Of Mice and Elephants: A Matter of Scale," *New York Times,* January 12, 1999. For a detailed analysis see John K.-J. Li, "Scaling and Invariants in Cardiovascular Biology," in Brown and West, *Scaling in Biology,* 113–22.

18 **sensible that mice might get more cancer:** Naked mole rats, however, appear to never succumb, possibly because of their ability to lower their metabolisms. They also live nine times longer than mice. See Sitai Liang et al., "Resistance to Experimental Tumorigenesis in Cells of a Long-lived Mammal, the Naked

Mole-rat," *Aging Cell* 9, no. 4 (August 2010): 626–35. For a popular account by two researchers see Thomas J. Park and Rochelle Buffenstein, "Underground Supermodels," *The Scientist,* June 1, 2012. Daniel Engber wrote about naked mole rats and cancer in "The Anti-Mouse," *Slate,* November 18, 2011.

18 **Scientists have proposed several reasons:** See, for example, Anders Bredberg, "Cancer Resistance and Peto's Paradox," *Proceedings of the National Academy of Sciences* 106, no. 20 (May 19, 2009): E51; and George Klein, "Reply to Bredberg: The Voice of the Whale," on page E52.

19 **hypertumors:** Nagy, Victor, and Cropper, "Why Don't All Whales Have Cancer?"

19 **thought there must be some kind of connection:** F. Galis, "Why Do Almost All Mammals Have Seven Cervical Vertebrae? Developmental Constraints, Hox Genes, and Cancer," *The Journal of Experimental Zoology* 285, no. 1 (April 15, 1999): 19–26.

20 **"Natural Compounds in Pomegranates":** These titles are from AACR press releases, American Association for Cancer Research website.

## CHAPTER 2 Nancy's Story

22 **two-thirds of cancer cases are preventable:** See, for example, World Cancer Research Fund/American Institute for Cancer Research, *Food, Nutrition, Physical Activity, and the Prevention of Cancer: A Global Perspective* (Washington, DC: AICR, 2007), xxv.

22 **the argument is weak at best:** Miguel A. Sanjoaquin et al., "Folate Intake and Colorectal Cancer Risk: A Meta-analytical Approach," *International Journal of Cancer* 113, no. 5 (February 20, 2005): 825–28; Susanna C. Larsson, Edward Giovannucci, and Alicja Wolk, "Folate and Risk of Breast Cancer: A Meta-analysis," *Journal of the National Cancer Institute* 99, no. 1 (January 3, 2007): 64–76; and Jane C. Figueiredo et al., "Folic Acid and Risk of Prostate Cancer: Results from a Randomized Clinical Trial," *Journal of the National Cancer Institute* 101, no. 6 (March 18, 2009): 432–35.

23 **folic acid . . . can increase cancer risk:** See, for example, Figueiredo et al., "Folic Acid and Risk of Prostate Cancer"; and Marta Ebbing et al., "Cancer Incidence and Mortality After Treatment with Folic Acid and Vitamin B12," *JAMA: The Journal of the American Medical Association* 302, no. 19 (November 18, 2009): 2119–26.

23 **administering antifolates:** John J. McGuire, "Anticancer Antifolates: Current Status and Future Directions," *Current Pharmaceutical Design* 9, no. 31 (2003): 2593–613.

23 **among the oldest chemotherapeutic drugs:** The pioneer in this research was Sidney Farber. See S. Farber et al., "Temporary Remissions in Acute Leukemia in Children Produced by Folic Acid Antagonist, 4-Aminopteroyl-Glutamic Acid," *New England Journal of Medicine* 238, no. 23 (June 3, 1948): 787–93. The story is told in Siddhartha Mukherjee's fine book *The Emperor of All Maladies: A Biography of Cancer* (New York: Scribner, 2010), 27–36.

23 **the mythology surrounding antioxidants:** Rudolf I. Salganik, "The Benefits and Hazards of Antioxidants," *Journal of the American College of Nutrition* 20 (2001): 464S–72S.

23 **a clinical trial in Finland:** "The Effect of Vitamin E and Beta Carotene on the Incidence of Lung Cancer and Other Cancers in Male Smokers," *New England Journal of Medicine* 330, no. 15 (April 14, 1994): 1029–35.

23 **A similar trial in the United States:** Gary E. Goodman et al., "The Beta-Carotene and Retinol Efficacy Trial," *Journal of the National Cancer Institute* 96, no. 23 (December 1, 2004): 1743–50.

23 **phytochemicals:** Lee W. Wattenberg, "Chemoprophylaxis of Carcinogenesis: A Review," part 1, *Cancer Research* 26, no. 7 (July 1, 1966): 1520–26.

24 **the evidence here is also meager:** A randomized controlled study of male doctors recently reported an annual cancer incidence of 1.7 percent among the multivitamin takers compared with 1.8 percent for the placebo group: J. Gaziano et al., "Multivitamins in the Prevention of Cancer in Men," *JAMA: The Journal of the American Medical Association* (published online October 17, 2012): 1–10. See the comment section of the paper for references to other studies finding neutral and even negative effects.

24 **5 A Day program:** "5 A Day for Better Health Program Evaluation Report: Executive Summary," National Cancer Institute website, last updated March 1, 2006.

24 **The evidence, alas:** Walter C. Willett, "Fruits, Vegetables, and Cancer Prevention: Turmoil in the Produce Section," *Journal of the National Cancer Institute* 102, no. 8 (April 21, 2010): 510–11.

24 **those most likely to volunteer:** Willett, "Fruits, Vegetables, and Cancer Prevention."

25 **The largest prospective study on diet and health:** The European Prospective Investigation into Cancer and Nutrition, or EPIC, is described on the website of the International Agency for Research on Cancer. For a summary with citations of key EPIC findings, see "Diet and Cancer: the Evidence," Cancer Research UK website, updated September 25, 2009.

25 **a very weak effect:** Willett, "Fruits, Vegetables, and Cancer Prevention"; and Paolo Boffetta et al., "Fruit and Vegetable Intake and Overall Cancer Risk in the European Prospective Investigation into Cancer and Nutrition," *Journal of*

*the National Cancer Institute* 102, no. 8 (April 21, 2010): 529–37. No evidence was found that fruits and vegetables help ward off cancer of the breast (Carla H. van Gils et al., "Consumption of Vegetables and Fruits and Risk of Breast Cancer," *JAMA: The Journal of the American Medical Association* 293, no. 2 [January 12, 2005]: 183–93) or cancer of the prostate (Timothy J. Key et al., "Fruits and Vegetables and Prostate Cancer," *International Journal of Cancer* 109, no. 1 [March 2004]: 119–24).

25 **possible benefits with a few cancers:** See, for example, Anthony B. Miller et al., "Fruits and Vegetables and Lung Cancer," *International Journal of Cancer* 108, no. 2 (January 10, 2004): 269–76; and Heiner Boeing et al., "Intake of Fruits and Vegetables and Risk of Cancer of the Upper Aero-digestive Tract," *Cancer Causes & Control* 17, no. 7 (September 2006): 957–69.

25 **said to nurture a mix of bacteria:** Constantine Iosif Fotiadis et al., "Role of Probiotics, Prebiotics and Synbiotics in Chemoprevention for Colorectal Cancer," *World Journal of Gastroenterology* 14, no. 42 (November 14, 2008): 6453–57; and Janelle C. Arthur and Christian Jobin, "The Struggle Within: Microbial Influences on Colorectal Cancer," *Inflammatory Bowel Diseases* 17, no. 1 (January 2011): 396–409.

25 **The case for fiber may be a little stronger:** See Teresa Norat et al., "The Associations Between Food, Nutrition and Physical Activity and the Risk of Colorectal Cancer," which is available along with other recent EPIC findings on the World Cancer Research Fund's Diet and Cancer Report website. See "Continuous Update Project Report Summary. Food, Nutrition, Physical Activity, and the Prevention of Colorectal Cancer" (2011).

25 **the evidence has been controversial:** EPIC's positive findings were published as Sheila A. Bingham et al., "Dietary Fibre in Food and Protection Against Colorectal Cancer in the European Prospective Investigation into Cancer and Nutrition," *Lancet* 361, no. 9368 (May 3, 2003): 1496–1501. For conflicting results from the Nurses' Health Study, see Scott Gottlieb, "Fibre Does Not Protect Against Colon Cancer," *BMJ: British Medical Journal* 318, no. 7179 (January 30, 1999): 281; and C. S. Fuchs, W. C. Willett, et al., "Dietary Fiber and the Risk of Colorectal Cancer and Adenoma in Women," *New England Journal of Medicine* 340, no. 3 (January 21, 1999): 169–76.

25 **no evidence of a reduction in colorectal polyps:** Arthur Schatzkin et al., "Lack of Effect of a Low-Fat, High-Fiber Diet on the Recurrence of Colorectal Adenomas," *New England Journal of Medicine* 342, no. 16 (April 20, 2000): 1149–55. Similar controlled studies have also found no relationship. See, for example, D. S. Alberts et al., "Lack of Effect of a High-fiber Cereal Supplement on the Recurrence of Colorectal Adenomas," *New England Journal of Medicine* 342, no. 16 (April 20, 2000): 1156–62; and Shirley A. Beresford et

al., "Low-fat Dietary Pattern and Risk of Colorectal Cancer," *JAMA: The Journal of the American Medical Association* 295, no. 6 (February 8, 2006): 643–54.

25 **no effect on the recurrence of breast cancer:** John P. Pierce et al., "Influence of a Diet Very High in Vegetables, Fruit, and Fiber and Low in Fat on Prognosis Following Treatment for Breast Cancer," *JAMA: The Journal of the American Medical Association* 298, no. 3 (July 18, 2007): 289–98.

25 **brussels sprouts, cabbage:** B. N. Ames, M. Profet, and L. S. Gold, "Nature's Chemicals and Synthetic Chemicals: Comparative Toxicology," *Proceedings of the National Academy of Sciences* 87, no. 19 (October 1990): 7782–86; and Bruce N. Ames, "Dietary Carcinogens and Anticarcinogens," *Science* 221, no. 4617 (September 23, 1983): 1256–64.

26 **eating a lot of red meat:** The calculation is for a fifty-year-old. See Teresa Norat et al., "Meat, Fish, and Colorectal Cancer Risk," *Journal of the National Cancer Institute* 97, no. 12 (June 15, 2005): 906–16; and Doris S. M. Chan et al., "Red and Processed Meat and Colorectal Cancer Incidence: Meta-Analysis of Prospective Studies," *PLOS ONE* 6, no. 6 (June 6, 2011).

26 **from 1.28 percent to 1.7 percent:** Norat et al., "Meat, Fish, and Colorectal Cancer Risk."

26 **fish, fish oils, and colon cancer prevention:** For evidence that eating fish discourages cancer by encouraging apoptosis and impeding cellular proliferation, see Youngmi Cho et al., "A Chemoprotective Fish Oil- and Pectin-Containing Diet Temporally Alters Gene Expression Profiles in Exfoliated Rat Colonocytes Throughout Oncogenesis," *Journal of Nutrition* 141, no. 6 (June 1, 2011): 1029–35. For another perspective, see Catherine H. MacLean et al., "Effects of Omega-3 Fatty Acids on Cancer Risk," *JAMA: The Journal of the American Medical Association* 295, no. 4 (January 25, 2006): 403–15.

26 **mammalian fat . . . has come under challenge:** Ross L. Prentice et al., "Low-Fat Dietary Pattern and Risk of Invasive Breast Cancer: The Women's Health Initiative Randomized Controlled Dietary Modification Trial," *JAMA: The Journal of the American Medical Association* 295, no. 6 (February 8, 2006): 629–42; and Shirley A. Beresford et al., "Low-Fat Dietary Pattern and Risk of Colorectal Cancer." For a summary, see "The Nutrition Source: Low-Fat Diet Not a Cure-All," Harvard School of Public Health website.

26 **sugar may pose a greater danger:** Gary Taubes, *Good Calories, Bad Calories: Fats, Carbs, and the Controversial Science of Diet and Health* (New York: Vintage, 2008); and Gary Taubes, *Why We Get Fat: And What to Do About It* (New York: Knopf, 2010).

26 **Obesity . . . has joined the short list:** See, for example, "AACR Cancer Progress Report," 2012, American Association for Cancer Research website.

26  **caloric restriction:** The mechanisms are complex, involving insulin regulation and other cellular processes. See Stephen D. Hursting et al., "Calorie Restriction, Aging, and Cancer Prevention," *Annual Review of Medicine* 54 (February 2003): 131–52; D. Kritchevsky, "Caloric Restriction and Cancer," *Journal of Nutritional Science and Vitaminology* 47, no. 1 (February 2001): 13–19; Sjoerd G. Elias et al., "Transient Caloric Restriction and Cancer Risk (The Netherlands)," *Cancer Causes & Control* 18, no. 1 (February 2007): 1–5; and David M. Klurfeld et al., "Reduction of Enhanced Mammary Carcinogenesis in LA/N-cp (Corpulent) Rats by Energy Restriction," *Proceedings of the Society for Experimental Biology and Medicine* 196, no. 4 (April 1, 1991): 381–84. In *Good Calories, Bad Calories* Taubes argues that the anticarcinogenic effects seen in the animal experiments come not from an overall reduction in calories but from limiting sugars and carbohydrates.

27  **a jolt of estrogen:** See, for example, Endogenous Hormones and Breast Cancer Collaborative Group, "Circulating Sex Hormones and Breast Cancer Risk Factors in Postmenopausal Women," *British Journal of Cancer* 105, no. 5 (2011): 709–22; A. Heather Eliassen et al., "Endogenous Steroid Hormone Concentrations and Risk of Breast Cancer Among Premenopausal Women," *Journal of the National Cancer Institute* 98, no. 19 (October 4, 2006): 1406–15; and Rudolf Kaaks et al., "Serum Sex Steroids in Premenopausal Women and Breast Cancer Risk," *Journal of the National Cancer Institute* 97, no. 10 (May 18, 2005): 755–65.

27  **the list of known human carcinogens:** National Toxicology Program, *Report on Carcinogens,* 12th ed. (Research Triangle Park, NC: U.S. Department of Health and Human Services, 2011). Available on the National Toxicology Program website.

27  **possibly increasing the risk of breast cancer:** See, for example, F. Clavel-Chapelon, "Differential Effects of Reproductive Factors on the Risk of Pre- and Postmenopausal Breast Cancer," *British Journal of Cancer* 86, no. 5 (March 4, 2002): 723–27.

27  **A few scientists blame the change on bisphenol A:** See, for example, Kembra L. Howdeshell et al., "Environmental Toxins: Exposure to Bisphenol A Advances Puberty," *Nature* 401, no. 6755 (October 21, 1999): 763–64; and Laura N. Vandenberg, Ana M. Soto, et al., "Bisphenol-A and the Great Divide: A Review of Controversies in the Field of Endocrine Disruption," *Endocrine Reviews* 30, no. 1 (February 1, 2009): 75–95.

27  **a more widely accepted explanation involves nutrition:** See Sandra Steingraber, "The Falling Age of Puberty in U.S. Girls," August 2007, Breast Cancer Fund website, which includes citations to the research; and Sarah E. Anderson, Gerard E. Dallal, and Aviva Must, "Relative Weight and Race In-

fluence Average Age at Menarche," part 1, *Pediatrics* 111, no. 4 (April 2003): 844–50.

27 **the age of menarche . . . has dropped:** Steingraber, "The Falling Age of Puberty," 20.

27 **Lactation also appears to hold estrogen in check:** See World Cancer Research Fund/American Institute for Cancer Research, *Food, Nutrition, Physical Activity, and the Prevention of Cancer,* 239–42.

27 **more menstrual cycles than her grandmother:** David Plotkin, "Good News and Bad News About Breast Cancer," *The Atlantic,* June 1998.

27 **Hormone therapies . . . associated with some cancers:** For an overview see two reports on the National Cancer Institute website: "Menopausal Hormone Therapy and Cancer" and "Diethylstilbestrol (DES) and Cancer," both reviewed December 5, 2011.

27 **obesity, especially in older women:** Sabina Rinaldi et al., "Anthropometric Measures, Endogenous Sex Steroids and Breast Cancer Risk in Postmenopausal Women," *International Journal of Cancer* 118, no. 11 (June 1, 2006): 2832–39; and Petra H. Lahmann et al., "A Prospective Study of Adiposity and Postmenopausal Breast Cancer Risk," *International Journal of Cancer* 103, no. 2 (November 4, 2002): 246–52.

27 **reduce the chances of premenopausal women getting breast cancer:** Kaaks et al., "Serum Sex Steroids in Premenopausal Women and Breast Cancer Risk." Also see Elisabete Weiderpass et al., "A Prospective Study of Body Size in Different Periods of Life and Risk of Premenopausal Breast Cancer," *Cancer Epidemiology, Biomarkers & Prevention* 13, no. 7 (July 2004): 1121–27; and L. J. Vatten and S. Kvinnsland, "Prospective Study of Height, Body Mass Index and Risk of Breast Cancer," *Acta Oncologica* 31, no. 2 (1992): 195–200.

27 **oral contraceptives may slightly raise the odds:** "Oral Contraceptives and Cancer Risk," National Cancer Institute, reviewed March 21, 2012.

28 **Alcohol . . . with digestive cancers:** The evidence for esophageal, liver, and other cancers is examined in Vincenzo Bagnardi et al., "Alcohol Consumption and the Risk of Cancer: A Meta-Analysis," *Alcohol Research and Health: The Journal of the National Institute on Alcohol Abuse and Alcoholism* 25, no. 4 (2001): 263–70.

28 **the risk from hepatitis viruses:** Heather M. Colvin and Abigail E. Mitchell, eds., *Hepatitis and Liver Cancer* (Washington, DC: The National Academies Press, 2010), 29–30.

28 **exposure to aflatoxin:** See, for example, P. E. Jackson and J. D. Groopman, "Aflatoxin and Liver Cancer," *Clinical Gastroenterology* 13, no. 4 (December 1999): 545–55.

28 **Consuming two or three drinks a day:** Wendy Y. Chen, Walter C. Willett, et al., "Moderate Alcohol Consumption During Adult Life, Drinking Patterns, and Breast Cancer Risk," *JAMA: The Journal of the American Medical Association* 306, no. 17 (November 2, 2011): 1884–90.

28 **a woman between the ages of forty and forty-nine:** "Risk of Developing Breast Cancer," Breastcancer.org website, last modified on March 14, 2012.

28 **Even tallness is a risk factor:** Jane Green et al., "Height and Cancer Incidence in the Million Women Study," *Lancet Oncology* 12, no. 8 (August 2011): 785–94.

28 **Ecuadoran villagers with a kind of dwarfism:** Jaime Guevara-Aguirre et al., "Growth Hormone Receptor Deficiency Is Associated with a Major Reduction in Pro-Aging Signaling, Cancer, and Diabetes in Humans," *Science Translational Medicine* 3, no. 70 (February 16, 2011): 70ra13; and Mitch Leslie, "Growth Defect Blocks Cancer and Diabetes," *Science* 331, no. 6019 (February 18, 2011): 837.

28 **measured not in small percentages:** See, for example, the rankings in the Harvard Cancer Risk Index, described in G. A. Colditz et al., "Harvard Report on Cancer Prevention Volume 4: Harvard Cancer Risk Index," *Cancer Causes & Control* 11, no. 6 (July 2000): 477–88.

28 **factors of ten to twenty:** According to the "Lung Cancer Fact Sheet," male smokers are twenty-three times more likely to develop lung cancer, and women thirteen times more likely, compared with people who never smoked (American Lung Association website, November 2010).

28 **the figure is more like 1 in 8:** Rebecca Goldin, "Lung Cancer Rates: What's Your Risk?" March 8, 2006, Research at Statistical Assessment Service (STATS) website, George Mason University.

29 **online Memorial Sloan-Kettering cancer prediction tool:** See "Cancer Care/Prediction Tools" on the Sloan-Kettering website.

29 **one-tenth the risk:** "Summary Report: Analysis of Exposure and Risks to the Public from Radionuclides and Chemicals Released by the Cerro Grande Fire at Los Alamos June 12, 2002," New Mexico Environment Department, Risk Assessment Corporation, report no. 5-NMED-2002-FINAL, Risk Assessment Corporation website.

30 **some evidence, weak and conflicting:** "Vitamin D and Cancer Prevention: Strengths and Limits of the Evidence," National Cancer Institute website, reviewed June 16, 2010; and Cindy D. Davis, "Vitamin D and Cancer: Current Dilemmas and Future Research Needs," *American Journal of Clinical Nutrition* 88, no. 2 (August 2008): 565S–69S.

30 **among male Finnish smokers:** Rachael Z. Stolzenberg-Solomon et al., "A Prospective Nested Case-control Study of Vitamin D Status and Pancreatic

Cancer Risk in Male Smokers," *Cancer Research* 66, no. 20 (October 15, 2006): 10213–19.

30 **a distant second place:** Office of Radiation and Indoor Air, *EPA Assessment of Risks from Radon in Homes* (Washington, DC: United States Environmental Protection Agency, June 2003), EPA website; and "WHO Handbook on Indoor Radon: a Public Health Perspective" (Geneva: World Health Organization, September 2009), WHO website.

30 **about 7 in 1,000:** Office of Radiation and Indoor Air, *EPA Assessment of Risks,* appendix D, 82. The number they give is, more precisely, 73 out of 10,000.

30 **constant exposure:** With some digging one can find that the calculations assume 70 percent of one's time is spent indoors. Office of Radiation and Indoor Air, *EPA Assessment of Risks,* 7, 44.

30 **an artist living there had reported:** Harry Otway and Jon Johnson, "A History of the Working Group to Address Los Alamos Community Health Concerns," Los Alamos National Laboratory, January 2000, available on the website of the U.S. Department of Energy, Office of Scientific and Technical Information.

31 **State health officials investigated:** William F. Athas and Charles R. Key, "Los Alamos Cancer Rate Study: Phase I," New Mexico Department of Health and New Mexico Tumor Registry, University of New Mexico Cancer Center, March 1993 (published on the UNM Health Sciences Center website); and William F. Athas, "Investigation of Excess Thyroid Cancer Incidence in Los Alamos County," Division of Epidemiology, Evaluation, and Planning, New Mexico Department of Health, April 1996. Available on the Department of Energy website.

31 **Texas sharpshooter effect:** The term was coined in the mid 1970s by the epidemiologist Seymour Grufferman while he was investigating a reported Hodgkin's lymphoma cluster on Long Island, New York. E-mail to author, June 10, 2012. Also see S. Grufferman, "Clustering and Aggregation of Exposures in Hodgkin's Disease," *Cancer* 39 (1977): 1829–33; K. J. Rothman, "A Sobering Start for the Cluster Busters' Conference," *American Journal of Epidemiology* 132, no. 1 suppl. (July 1990): S6–13; and Atul Gawande, "The Cancer-Cluster Myth," *New Yorker,* February 8, 1999.

31 **no harmful exposures from chemical or radioactive contamination:** Agency for Toxic Substances and Disease Registry, "Public Health Assessment for Los Alamos National Laboratory," September 8, 2006, available on the website of the U.S. Department of Health and Human Services, Agency for Toxic Substances and Disease Registry.

31 **The Long Island cancer cluster:** "Report to the U.S. Congress: The Long Island Breast Cancer Study Project" (Washington, DC: Department of Health

and Human Services, November 2004). The findings are summarized in Deborah M. Winn, "Science and Society: The Long Island Breast Cancer Study Project," *Nature Reviews Cancer* 5, no. 12 (December 2005): 986–94; and Renee Twombly, "Long Island Study Finds No Link Between Pollutants and Breast Cancer," *Journal of the National Cancer Institute* 94, no. 18 (2002): 1348–51.

32 **"a type of population control":** Patricia Braus, "Why Does Cancer Cluster?" *American Demographics,* March 1996.

32 **the median age for diagnosis of breast cancer:** "SEER Stat Fact Sheets: Breast," National Cancer Institute, Surveillance, Epidemiology and End Results website.

33 **"there is an environmental connection":** Braus, "Why Does Cancer Cluster?"

38 **reading about how this might have happened:** My first stop was Robert A. Weinberg, "How Cancer Arises," *Scientific American* 275, no. 3 (September 1996): 62–70.

CHAPTER 3   The Consolations of Anthropology

41 **When Louis Leakey sat down to recount:** He gave at least three versions of the story: L. S. B. Leakey, *The Stone Age Races of Kenya* (London: Oxford University Press, 1935), 10–11; Leakey, *By the Evidence* (New York: Harcourt Brace Jovanovich, 1974), 20–22, 35–36; and Leakey, *Adam's Ancestors* (London: Methuen & Co., 1934), 202–3. I also referred to Virginia Morell, *Ancestral Passions: The Leakey Family and the Quest for Humankind's Beginnings* (New York: Simon & Schuster, 1995), 65–71, 80–93.

42 **deposited . . . in Early Pleistocene time:** Morell, *Ancestral Passions,* 85.

42 **"not only the oldest known human fragment":** Leakey, *Stone Age Races,* 9.

42 **Java man and Peking man:** Piltdown man had not yet been exposed as a hoax.

42 **One of his detractors thought:** P. G. H. Boswell, "Human Remains from Kanam and Kanjera, Kenya Colony," *Nature* 135, no. 3410 (March 9, 1935): 371. Morell describes the controversy, including some bungling of the evidence by Leakey, in *Ancestral Passions,* 69, 80–93. For an excoriating interpretation of the event see Martin Pickford, *Louis S. B. Leakey: Beyond the Evidence* (London: Janus Publishing Company, 1997). Pickford and the Leakey family have been bitter enemies (Declan Butler, "The Battle of Tugen Hills," *Nature* 410, no. 6828 [March 29, 2001]: 508–9), and it can be difficult to separate the science from the politics. Pickford is also coauthor (with Eustace Gitonga) of a

book about Louis Leakey's son entitled *Richard E. Leakey: Master of Deceit* (Nairobi: White Elephant Publishers, 1995).

42 **a more distant relative like *Australopithecus*:** Kenneth P. Oakley, "The Kanam Jaw," *Nature* 185, no. 4717 (March 26, 1960): 945–46.

42 **Neanderthal man:** Phillip V. Tobias, "The Kanam Jaw," *Nature* 185, no. 4717 (March 26, 1960): 946–47.

42 **or *Homo habilis*:** That was the assessment of Harvard anthropologist David Pilbeam, who told Morell that the fossil may be as much as 2 or more million years old (*Ancestral Passions*, 9, note 11). He reconfirmed that in an e-mail to me, April 30, 2012.

42 **others have come to believe:** In "A Reconsideration of the Date of the Kanam Jaw," *Journal of Archaeological Science* 2, no. 2 (June 1975): 151–52, Kenneth P. Oakley theorized that the fossil "may have been enclosed in a Middle Pleistocene surface limestone block which was down-faulted in a fissure penetrating the older Kanam Beds." The Berkeley anthropologist Tim White concluded that the jaw is probably Late Pleistocene. See Eric Delson et al., eds., *Encyclopedia of Human Evolution and Prehistory* (New York: Garland, 2000), 739.

42 **no more than about 700,000 years old:** E-mail to author, May 7, 2012, from Richard Potts, director of the Human Origins Program at the Smithsonian Institution's National Museum of Natural History in Washington.

42 **carefully cleaning the specimen:** Leakey, *By the Evidence*, 20–22.

42 **diagnosed it as sarcoma of the bone:** J. W. P. Lawrence, Esq., "A Note on the Pathology of the Kanam Mandible," in Leakey, *Stone Age Races of Kenya*, appendix A, 139.

43 **There was also a thin fracture:** For a description of Kanam man's anatomical details see Leakey, *Stone Age Races*, 19–23.

43 **impossible to tell what Kanam man's chin had been like:** M. F. Ashley Montagu, "The Chin of the Kanam Mandible," *American Anthropologist* 59, no. 2 (April 1, 1957): 335–38.

43 **Another anthropologist disagreed:** Tobias, "The Kanam Jaw."

43 **an entirely different cancer:** G. Stathopoulos, "Letter: Kanam Mandible's Tumour," *Lancet* 305, no. 7899 (January 18, 1975): 165.

43 **Others were not so certain:** A. T. Sandison, "Kanam Mandible's Tumour," *Lancet* 305, no. 7901 (February 1, 1975): 279.

43 **Brothwell concluded:** Don Brothwell and A. T. Sandison, *Diseases in Antiquity: A Survey of the Diseases, Injuries and Surgery of Early Populations* (Springfield, IL: Charles C. Thomas, 1967), 330.

44 **scanning the mandible with an electron microscope:** J. Phelan, T. G. Bromage, et al., "Diagnosis of the Pathology of the Kanam Mandible," *Oral Sur-*

*gery, Oral Medicine, Oral Pathology, Oral Radiology and Endodontology* 103, no. 4 (April 2007): e20.

44 **"bone run amok":** Timothy Bromage, e-mail message to author, July 1, 2010.

44 **"the giant sloth":** Details are from a sign at the museum. My visit there was in May 2011.

45 **Leakey had sliced through the mass:** He reported that "a section was cut through the mandible in the region of the first molar" (*Stone Age Races,* 2). He also mentions x-ray radiographs.

46 **a small group of Greek and Egyptian oncologists:** Spiro Retsas, ed., *Palaeo-Oncology: The Antiquity of Cancer,* 5th ed. (London: Farrand Press, 1986), 7–9.

46 **"As a crab is furnished with claws":** Alexander Haddow, "Historical Notes on Cancer from the MSS. of Louis Westenra Sambon," *Proceedings of the Royal Society of Medicine* 29, no. 9 (July 1936): 1015–28.

46 **"because it adheres with such obstinacy":** Haddow, "Historical Notes on Cancer," 24.

47 **"[I]t attaches itself to the body of a young crab":** Haddow, "Historical Notes on Cancer," 25.

47 **by placing a live crab on top of it:** Haddow, "Historical Notes on Cancer," 28.

47 **"With treatment they soon die":** Retsas, *Palaeo-Oncology,* 45.

48 **a category of growth called *"praeter naturam"*:** Erwin H. Ackerknecht, "Historical Notes on Cancer," *Medical History* 2, no. 2 (April 1958): 114–19.

48 **"a tumor malignant and indurated":** Retsas, *Palaeo-Oncology,* 46.

48 **"The early cancer we have cured":** Retsas, *Palaeo-Oncology,* 49.

48 **"When a cancer has lasted long":** L. Weiss, "Metastasis of Cancer: A Conceptual History from Antiquity to the 1990s; Part 2: Early Concepts of Cancer," *Cancer Metastasis Reviews* 19, nos. 3–4 (2000): i-xi, 205–17.

48 **diagnosed in people fifty-five or older:** "Cancer Facts & Figures 2012," American Cancer Society website.

48 **hovering around thirty or forty years:** For a discussion of the difficulties of estimating past longevity, see J. R. Wilmoth, "Demography of Longevity: Past, Present, and Future Trends," *Experimental Gerontology* 35, nos. 9–10 (December 2000): 1111–29.

48 **the Saxon skeleton whose tumorous femur:** Brothwell and Sandison, *Diseases in Antiquity,* 331 and 339, figure 11b.

49 **comb through the bones:** One in 100,000 people gets osteosarcoma. See Lisa Mirabello, Rebecca J. Troisi, and Sharon A. Savage, "Osteosarcoma Incidence and Survival Rates from 1973 to 2004: Data from the Surveillance, Epidemiology, and End Results Program," *Cancer* 115, no. 7 (April 1, 2009): 1531–43.

49 **Iron Age man in Switzerland and a fifth-century Visigoth from Spain:**

Edward C. Halperin, "Paleo-oncology: The Role of Ancient Remains in the Study of Cancer," *Perspectives in Biology and Medicine* 47, no. 1 (2004): 1–14; Brothwell and Sandison, *Diseases in Antiquity,* 331; and Arthur C. Aufderheide and Conrado Rodriguez-Martin, *The Cambridge Encyclopedia of Human Paleopathology* (Cambridge: Cambridge University Press, 1998), 379.

49 **a medieval cemetery in the Black Forest Mountains:** K. W. Alt et al., "Infant Osteosarcoma," *International Journal of Osteoarchaeology* 12, no. 6 (December 24, 2002): 442–48.

49 **"to die a painful death":** Alt et al., "Infant Osteosarcoma," 447.

50 **"The large size of the tumor":** Eugen Strouhal, "Ancient Egyptian Case of Carcinoma," *Bulletin of the New York Academy of Medicine* 54, no. 3 (March 1978): 290–302.

50 **Traces were found in the skull:** Kurt W. Alt and Claus-Peter Adler, "Multiple Myeloma in an Early Medieval Skeleton," *International Journal of Osteoarchaeology* 2, no. 3 (May 23, 2005): 205–9; and C. Cattaneo et al., "Immunological Diagnosis of Multiple Myeloma in a Medieval Bone," *International Journal of Osteoarchaeology* 4, no. 1 (May 27, 2005): 1–2.

50 **Most skeletal cancers by far come from metastases:** Tony Waldron, "What Was the Prevalence of Malignant Disease in the Past?" *International Journal of Osteoarchaeology* 6, no. 5 (December 1, 1996): 463–70.

50 **discovered in Egyptian tombs:** Eugen Strouhal, "Tumors in the Remains of Ancient Egyptians," *American Journal of Physical Anthropology* 45, no. 3 (November 1, 1976): 613–20.

50 **in a Portuguese necropolis:** S. Assis and S. Codinha, "Metastatic Carcinoma in a 14th–19th Century Skeleton from Constância (Portugal)," *International Journal of Osteoarchaeology* 20, no. 5 (September 1, 2010): 603–20.

50 **in the Tennessee River valley:** Maria Ostendorf Smith, "A Probable Case of Metastatic Carcinoma from the Late Prehistoric Eastern Tennessee River Valley," *International Journal of Osteoarchaeology* 12, no. 4 (July 1, 2002): 235–47.

50 **in a leper skeleton from a medieval cemetery:** Donald J. Ortner, Keith Manchester, and Frances Lee, "Metastatic Carcinoma in a Leper Skeleton from a Medieval Cemetery in Chichester, England," *International Journal of Osteoarchaeology* 1, no. 2 (June 1, 1991): 91–98.

50 **near the Tower of London:** M. Melikian, "A Case of Metastatic Carcinoma from 18th Century London," *International Journal of Osteoarchaeology* 16, no. 2 (March 1, 2006): 138–44.

50 **excavated a 2,700-year-old burial mound:** Details of the discovery are described on the website of the German Archaeological Institute: "Complete Excavation of the Kurgan Arzhan 2 including an Undisturbed Royal Grave (late 7th century B.C.)." More information is on the website of the State Hermitage

Museum in St. Petersburg: "Restoration and Reconstruction of the Arzhan-2 Complex of Artifacts." I described this and some other cases more briefly in "Trying to Estimate Cancer Rates in Ancient Times," *New York Times*, December 27, 2010.

51 **his skeleton was infested with tumors:** Michael Schultz et al., "Oldest Known Case of Metastasizing Prostate Carcinoma Diagnosed in the Skeleton of a 2,700-year-old Scythian King from Arzhan (Siberia, Russia)," *International Journal of Cancer* 121, no. 12 (December 15, 2007): 2591–95.

51 **the partially cremated pelvis of a first-century Roman:** G. Grévin, R. Lagier, and C. A. Baud, "Metastatic Carcinoma of Presumed Prostatic Origin in Cremated Bones from the First Century A.D.," *Virchows Archiv: An International Journal of Pathology* 431, no. 3 (September 1997): 211–14.

51 **a skeleton from a fourteenth-century graveyard:** T. Anderson, J. Wakely, and A. Carter, "Medieval Example of Metastatic Carcinoma: A Dry Bone, Radiological, and SEM Study," *American Journal of Physical Anthropology* 89, no. 3 (November 1992): 309–23.

51 **osteoblastic . . . osteolytic:** Waldron, "What Was the Prevalence?"

51 **show the strongest appetite:** Tony Waldron, *Palaeopathology* (Cambridge: Cambridge University Press, 2009), 185.

51 **A middle-aged woman with osteolytic lesions:** M. J. Allison et al., "Metastatic Tumor of Bone in a Tiahuanaco Female," *Bulletin of the New York Academy of Medicine* 56, no. 6 (1980): 581–87.

51 **a Late Holocene hunter-gather:** L. H. Luna et al., "A Case of Multiple Metastasis in Late Holocene Hunter-gatherers from the Argentine Pampean Region," *International Journal of Osteoarchaeology* 18, no. 5 (November 14, 2007): 492–506.

52 **Like 90 percent of human cancers:** "Cancer Overview," Stanford School of Medicine Cancer Institute website.

52 **For children . . . only a fraction of cancers are carcinomas:** "Disease Information," St. Jude Children's Research Hospital website.

52 **often spread first to the lung or liver:** "Metastatic Cancer," National Cancer Institute website, reviewed May 23, 2011. Prostate cancer is drawn to bone, but it would probably have been less frequent when life spans were shorter.

52 **"swellings" and "eatings":** See, for example, Margaret M. Olszewski, "Concepts of Cancer from Antiquity to the Nineteenth Century," *University of Toronto Medical Journal* 87, no. 3 (May 2010); and Retsas, *Palaeo-Oncology*, 36.

52 **A rectal carcinoma in a 1,600-year-old mummy:** A. Rosalie David and Michael R. Zimmerman, "Cancer: An Old Disease, a New Disease or Something in Between?" *Nature Reviews Cancer* 10, no. 10 (October 2010): 728–33.

52 **diagnosed with bladder cancer:** Michael R. Zimmerman and Arthur C.

Aufderheide, "Seven Mummies of the Dakhleh Oasis, Egypt: Seventeen Diagnoses," *Paleopathology Newsletter* 150 (June 2010): 16–23.

52 **on the face of a Chilean child:** David and Zimmerman, "Cancer: An Old Disease, a New Disease?"

52 **nine pre-Columbian Incan mummies:** Oscar B. Urteaga and George T. Pack, "On the Antiquity of Melanoma," *Cancer* 19, no. 5 (May 1, 1966): 607–10.

53 **To prepare a pharaoh:** Leonard Weiss, "Observations on the Antiquity of Cancer and Metastasis," *Cancer and Metastasis Reviews* 19, nos. 3–4 (December 2000): 193–204.

53 **embalmed tumors can survive:** M. R. Zimmerman, "An Experimental Study of Mummification Pertinent to the Antiquity of Cancer," *Cancer* 40 (1977): 1358–62. In an experiment, a liver taken from a patient with metastatic carcinoma was dried in an oven and then rehydrated. Zimmerman observed that "the features of cancer (large, dark staining and highly variable nuclei and invasion of surrounding tissue) are well preserved by mummification and that mummified tumors are actually better preserved than normal tissue." E-mail to author, November 11, 2010.

53 **Ferrante I of Aragon:** The king was also obese and his bones were infused with lead and zinc. See Gino Fornaciari et al., "K-ras Mutation in the Tumour of King Ferrante I of Aragon (1431–94) and Environmental Mutagens at the Aragonese Court of Naples," *International Journal of Osteoarchaeology* 9, no. 5 (October 6, 1999): 302–6; Antonio Marchetti, Gino Fornaciari, et al., "K-RAS Mutation in the Tumour of Ferrante I of Aragon, King of Naples," *Lancet* 347, no. 9010 (May 1996): 1272; and Laura Ottini, Gino Fornaciari, et al., "Gene-Environment Interactions in the Pre-Industrial Era: The Cancer of King Ferrante I of Aragon (1431–1494)," *Human Pathology* 42, no. 3 (March 2011): 332–39.

53 **counted about two hundred suspected cancer sightings:** I started with 176 examples Strouhal had tabulated in what he called the Old World (reference in A. Sefcáková, E. Strouhal, et al., "Case of Metastatic Carcinoma from End of the 8th–early 9th Century Slovakia," *American Journal of Physical Anthropology* 116, no. 3 [November 2001]: 216–29) and then added in New World cases and cases found since the paper was published.

53 **stumbled on by chance:** Strouhal comments on this in "Tumors in the Remains of Ancient Egyptians."

53 **taphonomic changes:** Waldron, *Palaeopathology*, 21–23; Weiss, "Observations on the Antiquity of Cancer and Metastasis"; and E. Strouhal, "Malignant Tumors in the Old World," *Paleopathology Newsletter* no. 85, suppl. (1994): 1–6.

53 **pseudopathology:** Aufderheide and Rodriguez-Martin, *Cambridge Encyclopedia of Human Paleopathology*, 11–18.

54 **significantly underreported:** In *Diseases in Antiquity*, Brothwell speculates that "the scarcity of tumours has been overemphasized in the past—a fact which in itself may have depressed some detailed searching." See his chapter "The Evidence for Neoplasms," 320–45. Also see Waldron, "What Was the Prevalence?"

54 **more likely to appear in certain bones:** Waldron, *Palaeopathology*, 185.

54 **Hoping to cut through the uncertainty:** Waldron, "What Was the Prevalence?"

54 **between 0 and 2 percent for males and 4 and 7 percent for females:** See figure 1 of Waldron, "What Was the Prevalence?" The numbers were higher for women because of uterine and breast cancer. In the next century cancer in men would come to dominate because of cigarettes and lung cancer.

54 **The next step:** Andreas G. Nerlich et al., "Malignant Tumors in Two Ancient Populations: An Approach to Historical Tumor Epidemiology," *Oncology Reports* 16, no. 1 (July 2006): 197–202.

55 **an article that had just appeared:** David and Zimmerman, "Cancer: An Old Disease, a New Disease?"

55 **In a news release from her university:** "Scientists Suggest that Cancer Is Man-made," University of Manchester website, October 14, 2010.

56 **some take the number at face value:** See, for example, Luigi L. Capasso, "Antiquity of Cancer," *International Journal of Cancer* 113, no. 1 (January 1, 2005): 2–13; and M. S. Micozzi, "Diseases in Antiquity: The Case of Cancer," *Archives of Pathology and Laboratory Medicine* 115 (1991): 838–44.

56 **the total number of ancient and prehistoric skeletons:** The anthropologists I asked were Anne L. Grauer, president of the Paleopathology Association and an anthropologist at Loyola University in Chicago, Heather J. H. Edgar, curator of human osteology at the Maxwell Museum of Anthropology at the University of New Mexico, and Tim D. White, Professor of Integrative Biology at the University of California, Berkeley.

57 **A demographer . . . made a rough calculation:** Carl Haub, "How Many People Have Ever Lived on Earth?" October 2011, Population Reference Bureau website.

CHAPTER 4   Invasion of the Body Snatchers

58 **"rheumatism and debility":** T. R. Ashworth, "A Case of Cancer in Which Cells Similar to Those in the Tumours Were Seen in the Blood After Death," *Australian Medical Journal* 14 (1869): 146–47.

59  **secreting "morbid juices":** L. Weiss, "Concepts of Metastasis," *Cancer and Metastasis Reviews* 19 (2000): 219–34, which is part 3 of a longer piece, "Metastasis of Cancer: A Conceptual History from Antiquity to the 1990s," 193–400. I also relied on two other articles by Weiss in the same issue: "Observations on the Antiquity of Cancer and Metastasis" (193–204) and "Early Concepts of Cancer" (205–17). Other sources on the history of the cellular idea of cancer include James Stuart Olson, *The History of Cancer: An Annotated Bibliography* (New York: Greenwood Press, 1989); Erwin H. Ackerknecht, "Historical Notes on Cancer," *Medical History* 2, no. 2 (April 1958): 114–19; Margaret M. Olszewski, "Concepts of Cancer from Antiquity to the Nineteenth Century," *University of Toronto Medical Journal* 87, no. 3 (May 2010); and W. I. B. Onuigbo, "The Paradox of Virchow's Views on Cancer Metastasis," *Bulletin of the History of Medicine* 34 (1962): 444–49. Another valuable resource was Jacob Wolff, *The Science of Cancerous Disease from Earliest Times to the Present*, first published in 1907. It was translated from German by Barbara Ayoub and reissued in 1989 by Science History Publications and the National Library of Medicine.

59  **"metastatic affections":** Weiss, "Early Concepts of Cancer."

59  **an idea that was carried by Galen:** Ackerknecht, "Historical Notes."

59  **Descartes saw a connection:** Ackerknecht, "Historical Notes."

59  **A Parisian surgeon:** Weiss, "Concepts of Metastasis."

60  **traveling along the lymph vessel walls:** Weiss, "Concepts of Metastasis."

60  **Even the nervous system:** Weiss, "Concepts of Metastasis."

60  **leprosy and elephantiasis:** Ackerknecht, "Historical Notes."

60  **"cancer juice":** Ackerknecht, "Historical Notes."

60  **not sharp enough to show:** Wolff, *Science of Cancerous Disease*, 101–3.

60  **a book published in 1838:** There is an English version translated by Charles West as *On the Nature and Structural Characteristics of Cancer, and Those Morbid Growths Which May Be Confounded with It* (London: Sherwood, Gilbert, and Piper, 1840). For an excerpt see Johannes Müller, "On the Nature and Structural Characteristics of Cancer: General Observations on the Minute Structure of Morbid Growths," *CA: A Cancer Journal for Clinicians* 23, no. 5 (December 30, 2008): 307–12.

60  **from a primitive fluid called blastema:** Müller's ideas are summarized in Wolff, *Science of Cancerous Disease*, 108; and Olszewski, "Concepts of Cancer."

60  **Virchow, took the next step:** Ackerknecht, "Historical Notes."

60  **"a dissemination of cells":** Onuigbo, "The Paradox of Virchow's Views."

60  **all cancer arose from connective tissue:** Ackerknecht, "Historical Notes."

60  **Thiersch helped discredit that idea:** Ackerknecht, "Historical Notes."

61  **"Cancer is incurable":** Quoted in Weiss, "Early Concepts of Cancer."

61 **still not entirely clear today:** Robert A. Weinberg, *The Biology of Cancer* (New York: Garland Science, 2007), 593–94.

61 **encompass 3,914 pages:** Wolff, *Science of Cancerous Disease,* ix.

61 **"may or may not wish to compare":** The introduction is by the medical historian Saul Jarcho, MD.

62 **"When a plant goes to seed":** S. Paget, "The Distribution of Secondary Growths in Cancer of the Breast," *Lancet* 133, no. 3421 (1889): 571–73. It was republished as "Stephen Paget's Paper Reproduced from The Lancet, 1889," *Cancer and Metastasis Reviews* 8, no. 2 (1989): 98–101.

62 **it would soon be swamped:** Weinberg, *Biology of Cancer,* 636.

62 **head straight for the brain:** "Metastatic Brain Tumor," published online by the National Library of Medicine, Medline Plus website.

62 **Ian Hart and Isaiah Fidler:** Their paper is "Role of Organ Selectivity in the Determination of Metastatic Patterns of the B16 Melanoma," *Cancer Research* 40 (1980): 2281–87. Also see Isaiah J. Fidler, "The Pathogenesis of Cancer Metastasis: The 'Seed and Soil' Hypothesis Revisited," *Nature Reviews Cancer* 3, no. 6 (June 2003): 453–58.

62 **A video I came across:** "Overview of Metastasis," published online by CancerQuest, Winship Cancer Institute website, Emory University.

63 **The process is called anoikis:** Lance A. Liotta and Elise Kohn. "Anoikis: Cancer and the Homeless Cell," *Nature* 430, no. 7003 (August 26, 2004): 973–74.

63 **most will perish immediately:** For a fascinating account of the intricacies of metastasis, see Weinberg, *Biology of Cancer,* chapter 14. I also referred to Ann F. Chambers, Alan C. Groom, and Ian C. MacDonald, "Metastasis: Dissemination and Growth of Cancer Cells in Metastatic Sites," *Nature Reviews Cancer* 2, no. 8 (August 1, 2002): 563–72; and Christine L. Chaffer and Robert A. Weinberg, "A Perspective on Cancer Cell Metastasis," *Science* 331, no. 6024 (March 25, 2011): 1559–64.

63 **jettison enough of their cytoplasm:** Weinberg, *Biology of Cancer,* 593–94. He suggests a more likely explanation is that cancer cells can avoid the capillary trap by passing instead through arterial-venous shunts.

63 **researchers found that after twenty-four hours:** For a review see Fidler, "Pathogenesis of Cancer Metastasis."

63 **cancer in one breast:** Weinberg, *Biology of Cancer,* 636, sidebar 14.8.

64 **a molecular "zip code" identifying the organ:** Weinberg, *Biology of Cancer,* 637.

64 **priming them for survival:** Andy J. Minn, Joan Massagué, et al., "Genes That Mediate Breast Cancer Metastasis to Lung," *Nature* 436, no. 7050 (July 28, 2005): 518–24; and Paula D. Bos, J. Massagué, et al., "Genes That Mediate

Breast Cancer Metastasis to the Brain," *Nature* 459, no. 7249 (June 18, 2009): 1005–9.

64 **a premetastatic niche:** Rosandra N. Kaplan, Shahin Rafii, and David Lyden, "Preparing the 'Soil': The Premetastatic Niche," *Cancer Research* 66, no. 23 (December 1, 2006): 11089–93.

64 **the travelers can bring their own soil:** Dan G. Duda et al., "Malignant Cells Facilitate Lung Metastasis by Bringing Their Own Soil," *Proceedings of the National Academy of Sciences* 107, no. 50 (December 14, 2010): 21677–82.

64 **exchange signals with the natives:** The process is described in the general references on metastasis listed above.

64 **rejoin the battle at home:** Larry Norton and Joan Massagué, "Is Cancer a Disease of Self-seeding?" *Nature Medicine* 12, no. 8 (August 2006): 875–78; Mi-Young Kim, Joan Massagué, et al., "Tumor Self-seeding by Circulating Cancer Cells," *Cell* 139, no. 7 (December 24, 2009): 1315–26; and Elizabeth Comen, Larry Norton, and Joan Massagué, "Clinical Implications of Cancer Self-seeding," *Nature Reviews Clinical Oncology* 8, no. 6 (June 2011): 369–77.

65 **the ability to initiate angiogenesis:** J. Folkman et al., "Isolation of a Tumor Factor Responsible for Angiogenesis," *The Journal of Experimental Medicine* 133, no. 2 (February 1, 1971): 275–88.

65 **creating connections to the lymphatic system:** Viviane Mumprecht and Michael Detmar, "Lymphangiogenesis and Cancer Metastasis," *Journal of Cellular and Molecular Medicine* 13, no. 8A (August 2009): 1405–16.

65 **signals to a nearby lymph node:** Satoshi Hirakawa et al., "VEGF-C-induced Lymphangiogenesis in Sentinel Lymph Nodes Promotes Tumor Metastasis to Distant Sites," *Blood* 109, no. 3 (February 1, 2007): 1010–17.

66 **survival rate can be as high as 90 percent:** "Endometrial (Uterine) Cancer: Survival by Stage" and "How Is Endometrial Cancer Staged?" Both are on the American Cancer Society website, last revised July 25, 2012.

67 **Beauty Beyond Belief:** Packaged and sold by BBB Seed, Boulder, Colorado.

CHAPTER 5 Information Sickness

70 **experimenting with fruit flies:** H. J. Muller, "Artificial Transmutation of the Gene," *Science* 66, no. 1699 (July 22, 1927): 84–87.

70 **discovered in his monastery garden:** An English translation of Gregor Mendel's landmark paper, "Experiments in Plant Hybridization" (1865), can be found online at MendelWeb.

71 **That kind of clarity:** The experiments by Avery, Hershey, and Chase, and the discovery of DNA's double-helical structure, are described in Horace Freeland

Judson's *The Eighth Day of Creation: Makers of the Revolution in Biology*, expanded ed. (Cold Spring Harbor, NY: Cold Spring Harbor Laboratory Press, 1996). The seminal papers include Oswald T. Avery, Colin M. MacLeod, and Maclyn McCarty, "Studies on the Chemical Nature of the Substance Inducing Transformation of Pneumococcal Types," *The Journal of Experimental Medicine* 79, no. 2 (February 1, 1944): 137–58; A. D. Hershey and M. Chase, "Independent Functions of Viral Protein and Nucleic Acid in Growth of Bacteriophage," *The Journal of General Physiology* 36, no. 1 (May 1952): 39–56; and J. D. Watson and F. H. C. Crick, "A Structure for Deoxyribose Nucleic Acid," *Nature* 171 (1953): 737–38. An annotated version of Watson and Crick's paper can be found on the website for the Exploratorium. See "Origins, Unwinding DNA, Life at Cold Spring Harbor Laboratory."

71 **x-rays were first produced:** For a translation of the original paper see W. C. Röntgen, "On a New Kind of Rays" (1895), republished in Wilhelm Conrad Röntgen, Sir George Gabriel Stokes, and Sir Joseph John Thomson, *Röntgen Rays: Memoirs by Röntgen, Stokes, and J. J. Thomson* (New York: Harper & Brothers, 1899), 3–13. The collection also includes Röntgen's second and third communications. Like the Curies, he had no reason yet to be fearful of ionizing radiation. He matter-of-factly describes what happens when he shines x-rays into his eyes (pp. 7 and 39–40).

71 **strange-looking chromosomes:** For Boveri's speculations about cancer cells, see "Concerning the Origin of Malignant Tumours," a translation by Henry Harris of Boveri's *Zur Frage der Entstehung maligner Tumoren* (1914), *Journal of Cell Science* 121 (January 1, 2008): 1–84. It has also been published as a book: Theodor Boveri, *Concerning the Origin of Malignant Tumours*, 1st ed. (Cold Spring Harbor, NY: Cold Spring Harbor Laboratory Press, 2007).

72 **to "multiply without restraint":** Boveri, "Concerning the Origin."

72 **"conceivable at least that mammalian cancer":** Volker Wunderlich, "Early References to the Mutational Origin of Cancer," *International Journal of Epidemiology* 36, no. 1 (February 1, 2007): 246–47.

72 **"a new kind of cell":** Wunderlich, "Early References."

72 **Becquerel accidentally discovered:** "On Radioactivity, a New Property of Matter," *Nobel Lectures, Physics 1901–1921* (Amsterdam: Elsevier Publishing Company, 1967), 52–70. This lecture, delivered on December 11, 1903, is available on the Nobel Prize website.

72 **Marie Curie noticed:** The Curies' experiments are described in Pierre Curie's June 6, 1905 Nobel lecture, "Radioactive Substances, Especially Radium," in *Nobel Lectures, Physics 1901–1921* (Amsterdam: Elsevier Publishing Company, 1967). Available at the Nobel Prize website. Also see Eve Curie, *Madame Curie: A Biography*, trans. Vincent Sheean (Garden City, NY: Doubleday,

Doran & Co., 1937); and Barbara Goldsmith, *Obsessive Genius: The Inner World of Marie Curie* (New York: W. W. Norton, 2005).

73 **"a kind of matter in the world":** The film with Greer Garson and Walter Pidgeon was nominated for the 1944 Academy Award for Outstanding Motion Picture. (The winner was *Casablanca*.)

73 **"One of our joys":** Marie Curie, *Pierre Curie (With the Autobiographical Notes of Marie Curie)*, trans. Charlotte Kellogg (New York: Macmillan Co., 1923), 187.

73 **an optical analog of a sonic boom:** More specifically the Curies were seeing Cherenkov radiation.

73 **decorate their teeth, fingernails, and eyebrows:** For reports on the Radium Girls, see Frederick L. Hoffman, "Radium (Mesothorium) Necrosis," *Journal of the American Medical Association* 85, no. 13 (1925): 961–65; R. E. Rowland, Radium in Humans: A Review of U.S. Studies, Argonne National Laboratory, Environmental Research Division, 1994; and Ross Mullner, *Deadly Glow: The Radium Dial Worker Tragedy* (Washington, DC: American Public Health Association, 1999).

74 **"soot warts":** "Cancer Scroti," in *The Chirurgical Works of Percival Pott,* vol. 3 (London: Johnson, 1808), 177–80.

74 **The same cancer was later found:** H. A. Waldron, "A Brief History of Scrotal Cancer," *British Journal of Industrial Medicine* 40, no. 4 (November 1983): 390–401.

74 **applying coal tar to rabbits' ears:** K. Yamagiwa and K. Ichikawa, "Experimental Study of the Pathogenesis of Carcinoma," *Journal of Cancer Research* 3 (1918): 1–29. Republished in *CA: A Cancer Journal for Clinicians* 27, no. 3 (December 31, 2008): 174–81.

74 **produced tumors in laboratory animals:** See, for example, J. W. Cook, C. L. Hewett, and I. Hieger, "The Isolation of a Cancer-producing Hydrocarbon from Coal Tar," *Journal of the Chemical Society* (January 1, 1933): 395–405.

75 **the Ames test:** Bruce N. Ames et al., "Carcinogens Are Mutagens: A Simple Test System Combining Liver Homogenates for Activation and Bacteria for Detection," *Proceedings of the National Academy of Sciences* 70, no. 8 (August 1973): 2281–85.

75 **studying chicken tumors:** Peyton Rous's papers are "A Transmissible Avian Neoplasm," *Journal of Experimental Medicine* 12, no. 5 (September 1, 1910): 696–705 and "A Sarcoma of the Fowl Transmissible by an Agent Separable from the Tumor Cells," *Journal of Experimental Medicine* 13, no. 4 (April 1, 1911): 397–411.

76 ***src, ras, fes, myb, myc*:** The string of revelations, which has been described as

the Revolution of 1976, was set off by Harold Varmus and J. Michael Bishop (D. Stehelin, H. E. Varmus, J. M. Bishop, and P. K. Vogt, "DNA Related to the Transforming Gene(s) of Avian Sarcoma Viruses Is Present in Normal Avian DNA," *Nature* 260, no. 5547 [March 11, 1976]: 170–73) and is described in Robert Weinberg's *One Renegade Cell: The Quest for the Origin of Cancer* (New York: Basic Books, 1999). I also referred to Weinberg's "How Cancer Arises," *Scientific American* 275, no. 3 (September 1996): 62–70; Douglas Hanahan and R. A. Weinberg, "The Hallmarks of Cancer," *Cell* 100, no. 1 (January 7, 2000): 57–70; and D. Hanahan and R. A. Weinberg, "Hallmarks of Cancer: The Next Generation," *Cell* 144, no. 5 (March 4, 2011): 646–74. Natalie Angier told Weinberg's story in *Natural Obsessions: Striving to Unlock the Deepest Secrets of the Cancer Cell* (New York: Warner Books, 1989), and Weinberg gave his own account in *Racing to the Beginning of the Road: The Search for the Origin of Cancer* (New York: Harmony, 1996).

77 **they were named proto-oncogenes:** C. Shih, R. A. Weinberg, et al., "Passage of Phenotypes of Chemically Transformed Cells via Transfection of DNA and Chromatin," *Proceedings of the National Academy of Sciences* 76, no. 11 (November 1979): 5714–18; and C. J. Tabin, R. A. Weinberg, et al., "Mechanism of Activation of a Human Oncogene," *Nature* 300, no. 5888 (November 11, 1982): 143–49.

78 **Some mutations are even more wrenching:** The best known example is the Philadelphia chromosome, which is involved in chronic myeloid leukemia. For the original report, see Peter Nowell and David Hungerford, "A Minute Chromosome in Chronic Granulocytic Leukemia," *Science* 132, no. 3438 (November 1960): 1497.

78 **when a gene called *Rb*:** S. H. Friend, R. A. Weinberg, et al., "A Human DNA Segment with Properties of the Gene That Predisposes to Retinoblastoma and Osteosarcoma," *Nature* 323, no. 6089 (October 16, 1986): 643–46; and J. A. DeCaprio et al., "The Product of the Retinoblastoma Susceptibility Gene Has Properties of a Cell Cycle Regulatory Element," *Cell* 58, no. 6 (September 22, 1989): 1085–95.

79 **both copies must be knocked out:** This is known as the two-hit hypothesis. See Alfred G. Knudson, "Mutation and Cancer: Statistical Study of Retinoblastoma," *Proceedings of the National Academy of Sciences* 68, no. 4 (April 1971): 820–23.

79 **involved in the timekeeping:** See, for example, DeCaprio et al., "The Product of the Retinoblastoma Susceptibility Gene."

79 **sits at the center of a web:** C. A. Finlay, P. W. Hinds, and A. J. Levine, "The P53 Proto-oncogene Can Act as a Suppressor of Transformation," *Cell* 57, no. 7 (June 30, 1989): 1083–93; and M. B. Kastan, B. Vogelstein, et al., "Participa-

tion of P53 Protein in the Cellular Response to DNA Damage," part 1, *Cancer Research* 51, no. 23 (December 1, 1991): 6304–11.

80 **programmed cell death, or apoptosis:** J. F. Kerr, A. H. Wyllie, and A. R. Currie, "Apoptosis: A Basic Biological Phenomenon with Wide-ranging Implications in Tissue Kinetics," *British Journal of Cancer* 26, no. 4 (August 1972): 239–57.

80 **a principle called the Hayflick limit:** L. Hayflick and P. S. Moorhead, "The Serial Cultivation of Human Diploid Cell Strains," *Experimental Cell Research* 25, no. 3 (December 1961): 585–621.

80 **The count is kept by telomeres:** The story of the discovery is told in Elizabeth H. Blackburn, Carol W. Greider, and Jack W. Szostak, "Telomeres and Telomerase: The Path from Maize, Tetrahymena and Yeast to Human Cancer and Aging," *Nature Medicine* 12, no. 10 (October 2006): 1133–38. The key papers are J. W. Szostak and E. H. Blackburn, "Cloning Yeast Telomeres on Linear Plasmid Vectors," *Cell* 29, no. 1 (May 1982): 245–55; C. W. Greider and E. H. Blackburn, "Identification of a Specific Telomere Terminal Transferase Activity in Tetrahymena Extracts, *Cell* 43(1985): 405–13; and C. W. Greider and E. H. Blackburn, "A Telomeric Sequence in the RNA of Tetrahymena Telomerase Required for Telomere Repeat Synthesis," *Nature* 337 (1989): 331–37.

80 **accumulating mutations:** Accelerating the process may be a phenomenon called genomic instability. See Simona Negrini, Vassilis G. Gorgoulis, and Thanos D. Halazonetis, "Genomic Instability—An Evolving Hallmark of Cancer," *Nature Reviews Molecular Cell Biology* 11, no. 3 (March 1, 2010): 220–28.

80 **As this evolution unfolds:** For an overview of the phenomenon, see Hanahan and Weinberg's "The Hallmarks of Cancer" and "Hallmarks of Cancer: The Next Generation."

80 **signals are sent to healthy cells:** These discoveries grew from early research on the role of the tumor microenvironment. See, for example, D. S. Dolberg and M. J. Bissell, "Inability of Rous Sarcoma Virus to Cause Sarcomas in the Avian Embryo," *Nature* 309, no. 5968 (June 7, 1984): 552–56; and D. S. Dolberg, M. J. Bissell, et al., "Wounding and Its Role in RSV-mediated Tumor Formation," *Science* 230, no. 4726 (November 8, 1985): 676–78.

81 **Macrophages and other inflammatory cells:** Lisa M. Coussens and Zena Werb, "Inflammation and Cancer," *Nature* 420, no. 6917 (December 19, 2002): 860–67.

81 **compared to bodily organs:** Mina J. Bissell and Derek Radisky, "Putting Tumours in Context," *Nature Reviews Cancer* 1, no. 1 (October 2001): 46–54.

CHAPTER 6 "How Heart Cells Embrace Their Fate"

82 **an embryo is so much like a tumor:** The complex process of implantation is described in Haibin Wang and Sudhansu K. Dey, "Roadmap to Embryo Implantation: Clues from Mouse Models," *Nature Reviews Genetics* 7, no. 3 (March 1, 2006): 185–99. For some of the parallels with tumorigenesis see Michael J. Murray and Bruce A. Lessey, "Embryo Implantation and Tumor Metastasis: Common Pathways of Invasion and Angiogenesis," *Seminars in Reproductive Medicine* 17, no. 3 (March 15, 2008): 275–90.

82 **enzymes erode the surface:** L. A. Salamonsen, "Role of Proteases in Implantation," *Reviews of Reproduction* 4, no. 1 (January 1999): 11–22.

82 **molecules help ensure a tight grip:** Maaike S. M. van Mourik et al., "Embryonic Implantation: Cytokines, Adhesion Molecules, and Immune Cells in Establishing an Implantation Environment," *Journal of Leukocyte Biology* 85, no. 1 (January 2009): 4–19.

82 **messages are sent to the immune system:** Van Mourik et al., "Embryonic Implantation."

82 **begins stimulating angiogenesis:** D. M. Sherer and O. Abulafia, "Angiogenesis During Implantation, and Placental and Early Embryonic Development," *Placenta* 22, no. 1 (January 2001): 1–13.

83 **the more parallels they find:** Melissa Marino, "In the Beginning: What Developmental Biology Can Teach About Cancer," *Lens* online magazine, Vanderbilt Medical Center website, February 2007.

83 **epithelial-mesenchymal transition:** The seminal article is Jean Paul Thiery, "Epithelial-Mesenchymal Transitions in Tumour Progression," *Nature Reviews Cancer* 2, no. 6 (June 2002): 442–54. Good reviews include Yibin Kang and Joan Massagué, "Epithelial-Mesenchymal Transitions: Twist in Development and Metastasis," *Cell* 118, no. 3 (August 6, 2004): 277–79; Jonathan M. Lee et al., "The Epithelial-Mesenchymal Transition: New Insights in Signaling, Development, and Disease," *Journal of Cell Biology* 172, no. 7 (March 27, 2006): 973–81; Jing Yang and Robert A. Weinberg, "Epithelial-Mesenchymal Transition: At the Crossroads of Development and Tumor Metastasis," *Developmental Cell* 14, no. 6 (June 2008): 818–29; and Raghu Kalluri and Robert A. Weinberg, "The Basics of Epithelial-Mesenchymal Transition," *Journal of Clinical Investigation* 119, no. 6 (June 1, 2009): 1420–28. For an account by some naysayers see David Tarin, Erik W. Thompson, and Donald F. Newgreen, "The Fallacy of Epithelial Mesenchymal Transition in Neoplasia," *Cancer Research* 65, no. 14 (July 15, 2005): 5996–6001. Both sides of the controversy are described in Heidi Ledford, "Cancer Theory Faces Doubts," *Nature* 472, no. 7343 (April 21, 2011): 273.

84 **holding its annual meeting:** Society for Developmental Biology Sixty-Ninth Annual Meeting, August 5–9, 2010, Albuquerque, NM. I also attended the Seventieth Annual Meeting, July 21– 22, 2011, in Chicago. For a nice overview of developmental biology, see Sean B. Carroll, *Endless Forms Most Beautiful: The New Science of Evo Devo* (New York: Norton, 2006). The website of the Society for Developmental Biology provides a portal to numerous resources like WormAtlas, with detailed and annotated maps of *C. elegans,* and Fly-Brain, which covers the *Drosophila* nervous system.

84 **"Organogenesis":** The proceedings of the Albuquerque conference are in *Developmental Biology* 344, no. 1 (2010): 391–542.

84 **wingless, frizzled, smoothened, patched, and disheveled:** Though I have tried to be consistent in my own usage, I have not religiously followed the rules for when to render the names and symbols of genes in upper case or lower case or italics or roman. Apologies to the specialists who may find that distracting.

85 **possible treatments for baldness:** Andrzej Dlugosz, "The Hedgehog and the Hair Follicle: A Growing Relationship," *Journal of Clinical Investigation* 104, no. 7 (October 1, 1999): 851–53.

85 **"The quirky sense of humour":** Ken Maclean, "Humour of Gene Names Lost in Translation to Patients," *Nature* 439, no. 7074 (January 19, 2006): 266.

85 **It now goes by the less evocative name Zbtb7:** Tom Simonite, "Pokemon Blocks Gene Name," *Nature* 438, no. 7070 (December 14, 2005): 897.

85 **Since it was discovered in 1993:** R. D. Riddle, C. Tabin, et al., "Sonic Hedgehog Mediates the Polarizing Activity of the ZPA," *Cell* 75, no. 7 (December 31, 1993): 1401–16.

85 **sheep grazing in the mountains:** The story of cyclopamine is told in Philipp Heretsch, Lito Tzagkaroulaki, and Athanassios Giannis, "Cyclopamine and Hedgehog Signaling: Chemistry, Biology, Medical Perspectives," *Angewandte Chemie* (international ed. in English) 49, no. 20 (May 3, 2010): 3418–27.

86 **holoprosencephaly:** Max Muenk, "Translational Concepts to Disease: Holoprosencephaly as an Example," lecture presented July 22, 2011 at the Society for Developmental Biology Seventieth Annual Meeting, Chicago, IL.

86 **one of every 250 early embryos:** Erich Roessler, Maximilian Muenke, et al., "Mutations in the Human Sonic Hedgehog Gene Cause Holoprosencephaly," *Nature Genetics* 14, no. 3 (November 1996): 357–60.

86 **can drive the formation of malignancies:** For an overview of sonic hedgehog and cancer see Lee L. Rubin and Frederic J. de Sauvage, "Targeting the Hedgehog Pathway in Cancer," *Nature Reviews Drug Discovery* 5, no. 12 (December 2006): 1026–33; and Jennifer M. Bailey, Pankaj K. Singh, and Michael A. Hollingsworth, "Cancer Metastasis Facilitated by Developmental

Pathways: Sonic Hedgehog, Notch, and Bone Morphogenic Proteins," *Journal of Cellular Biochemistry* 102, no. 4 (November 1, 2007): 829–39.

86 **Gorlin syndrome:** Ervin H. Epstein, "Basal Cell Carcinomas: Attack of the Hedgehog," *Nature Reviews Cancer* 8, no. 10 (October 2008): 743–54.

86 **a cream containing cyclopamine:** Sinan Tabs and Oktay Avci, "Induction of the Differentiation and Apoptosis of Tumor Cells in Vivo with Efficiency and Selectivity," *European Journal of Dermatology* 14, no. 2 (April 2004): 96–102.

87 **another hedgehog inhibitor:** "FDA Approval for Vismodegib," National Cancer Institute.

87 **might help control the genetic switches:** More specifically, Dmrt5 is a transcription factor, a molecule that binds to the genome and regulates the output of a gene.

87 **"zinc fingers," "Dmrt5," and "Emma Farley":** Emma K. Farley et al., "Novel Transcription Factor Involved in Neurogenesis," *Developmental Biology* 344, no. 1 (2010): 493.

88 **so many new scraps of information:** Venugopala Reddy Bommireddy Venkata, Cordelia Rauskolb, and Kenneth D. Irvine, "Fat-Hippo Signaling Regulates the Proliferation and Differentiation of Drosophila Optic Neuroepithelia," *Developmental Biology* 344, no. 1 (2010): 506; and Thomas L. Gallagher and Joshua Arribere, "Fox1 and Fox4 Regulate Muscle-specific Splicing in Zebrafish and Are Required for Cardiac and Skeletal Muscle Functions," *Developmental Biology* 344, no. 1 (2010): 491–92.

88 **a whimsical turn:** Cristina L. Walcher and Jennifer L. Nemhauser, "1 + 1 = 3: When Two Hormones Are Better Than One," *Developmental Biology* 344, no. 1 (2010): 487; and Nowlan Freese and Susan C. Chapman, "Where'd My Tail Go?" *Developmental Biology* 344, no. 1 (2010): 441.

88 **six little words:** The full title was "How Heart Cells Embrace Their Fate in the Chordate *Ciona Intestinalis*" by Stacia Ilchena and James Cooley, *Developmental Biology* 344, no. 1 (2010): 502–3.

CHAPTER 7 Where Cancer Really Comes From

92 **excavating a canal:** Love Canal's history is described in Allan Mazur, *A Hazardous Inquiry: The Rashomon Effect at Love Canal* (Cambridge, MA: Harvard University Press, 1998), 8–15.

93 **some 22,000 tons of toxic waste:** "Love Canal: A Special Report to the Governor & Legislature," April 1981, New York Department of Health website.

93 **"The odors penetrate your clothing":** "Chemical Waste at Love Canal, October 18, 1977," Ecumenical Task Force of the Niagara Frontier Records,

1946–1995 (MS 65), University at Buffalo Libraries, Love Canal Collections website.

93 **incorporated the saga into a novel:** Joyce Carol Oates, *The Falls: A Novel* (New York: Ecco, 2004).

93 **the EPA estimated:** "U.S. Finds Risk of Cancer High for Residents Near Love Canal; Three Families Inside Fence," *New York Times,* November 11, 1979.

94 **admitted to a mathematical error:** Irvin Molotsky, "Rep. LaFalce Says Draft Report Inflated Love Canal Cancer Risk; Mathematical Errors Conceded," *New York Times,* November 20, 1979.

94 **Another EPA report found:** Irvin Molotsky, "Damage to Chromosomes Found in Love Canal Tests," *New York Times,* May 17, 1980. The findings were in D. Picciano, "Pilot Cytogenetic Study of the Residents Living Near Love Canal, a Hazardous Waste Site, *Mammalian Chromosome Newsletter* 21 (1980): 86–93.

94 **dismissed by a panel:** Richard J. Meislin, "Carey Panel Discounts 2 Studies on Love Canal Health Problems," *New York Times,* October 11, 1980.

94 **A later study for the Centers for Disease Control:** Clark W. Heath Jr. et al., "Cytogenetic Findings in Persons Living Near the Love Canal," *JAMA: The Journal of the American Medical Association* 251, no. 11 (March 16, 1984): 1437–40.

94 **a thirty-year retrospective:** "Love Canal Follow-up Health Study," New York Department of Health website, October 2008. For a critique of the report see appendix T, "Public Comments and Responses, Love Canal Follow-up Health Study," especially the comments by epidemiologist Richard Clapp, 145–47.

94 **Almost half of the 6,026 residents:** Demographic details are described in table 20, page 97 of the "Love Canal Follow-up."

94 **the birth defect rate:** "Love Canal Follow-up," 42–43. Altogether there were thirty-seven congenital malformations reported among 1,767 singleton births (those not including twins and triplets) between 1960 and 1996. For comparing incidence rates, the study counted only "consistently and reliably reported" cases as defined by the New York State Congenital Malformation Registry, which had complete records only beginning in 1983. Earlier information came from local hospitals and was not considered solid enough to use in the comparisons. (See pages 29–30 of the report for details. Also see table 19, page 96, and appendix A on page 103.)

94 **higher than for the rest of the state:** That excluded New York City.

94 **slightly more girls had been born:** "Love Canal Follow-up," 41–43.

94 **no convincing evidence:** "Love Canal Follow-up," 2.

94 **overall cancer rate was actually a little lower:** That was compared with both the county and the state. See "Love Canal Follow-up," 39–41.

95 **Rachel Carson's elegant warning:** *Silent Spring* (Boston: Houghton Mifflin Company, 1962).

95 **scathing polemics:** Samuel S. Epstein, *The Politics of Cancer* (San Francisco: Sierra Club Books, 1978) and *The Politics of Cancer Revisited* (Fremont Center, NY: East Ridge Press, 1998).

95 **a modern epidemic of cancer:** The historian Robert N. Proctor captures the zeitgeist in *Cancer Wars: How Politics Shapes What We Know and Don't Know About Cancer* (New York: Basic Books, 1995). See particularly 54–74.

95 **"the plague of the twentieth century":** Epstein, *Politics of Cancer Revisited,* 14.

95 **"a grim game of chemical roulette":** Russell Train in a speech to the National Press Club, February 26, 1976. The story was picked up by the Associated Press and appeared the next day in newspapers, including *The Morning Record* (Meriden, CT) and the *Sarasota Herald-Tribune.*

95 **"the Great Cancer Wars":** Proctor, *Cancer Wars,* 74.

95 *Ninety percent of cancer is environmental:* For an account of the origin of the misconception, see Proctor, *Cancer Wars,* 55–57; and (keeping in mind its libertarian bias) Edith Efron, *Apocalyptics: Cancer and the Big Lie: How Environmental Politics Controls What We Know About Cancer* (New York: Simon & Schuster, 1984), 429–32.

95 **known and suspected agents:** See National Toxicology Program, *Report on Carcinogens,* 12th ed. (Research Triangle Park, NC: U.S. Department of Health and Human Services, 2011).

96 **To get a sense of how strongly cancer was influenced:** The migrant studies are summarized in R. Doll and R. Peto, "The Causes of Cancer: Quantitative Estimates of Avoidable Risks of Cancer in the United States Today," *Journal of the National Cancer Institute* 66, no. 6 (June 1981): 1191–1308, reference on 1200–01; Proctor, *Cancer Wars,* 24–26; and Efron, *Apocalyptics,* 430–32.

97 **seemed to be escalating abruptly:** The numbers were published in Earl S. Pollack and John W. Horm, "Trends in Cancer Incidence and Mortality in the United States, 1969–76," *Journal of the National Cancer Institute* 64, no. 5 (May 1, 1980): 1091–103; and in *Toxic Chemicals and Public Protection: A Report to the President by the Toxic Substances Strategy Committee* (Washington, DC: Council on Environmental Quality, May 1980), which is available online through the Hathi Trust Digital Library.

97 **appeared to be the vindication:** For a description of the controversy see Doll and Peto, "Causes of Cancer," 1279–81; and Efron, *Apocalyptics,* 434–36.

97 **warned that the comparisons were invalid:** Doll and Peto, "Causes of Cancer," 1280–81; and Efron, *Apocalyptics,* 435.

97 **commissioned a study:** Doll and Peto, "Causes of Cancer."

97 **accomplished scientists in their field:** Doll's later work was called into question when it was revealed after his death that he had accepted consulting fees from chemical companies. In defending his colleague, Peto said that Doll was open about the connections and gave the money to Green College, Oxford, which he helped found. See Sarah Boseley, "Renowned Cancer Scientist Was Paid by Chemical Firm for 20 Years," *The Guardian*, December 7, 2006. In the letters section of the next day's edition other prominent scientists defended Doll's impartiality (see "Richard Doll Still Deserves Our Respect"). They included the chief executive of the Medical Research Council, the director of Wellcome Trust, and Martin Rees, the president of the Royal Society.

97 **which numbers to trust:** Doll and Peto, "Causes of Cancer," appendix C, 1270–81.

98 **The situation improved:** Doll and Peto, "Causes of Cancer," 1281.

98 **In 30 percent of cancer deaths, tobacco was a cause:** Doll and Peto, "Causes of Cancer," table 20, 1256.

99 **"most of the types of cancer that are common today":** Doll and Peto, "Causes of Cancer," 1212.

99 **Any specific case of cancer will have multiple causes:** For the dilemmas involved in sorting out the environmental and genetic factors of a disease, see Kenneth J. Rothman and Sander Greenland, "Causation and Causal Inference in Epidemiology," *American Journal of Public Health* 95 suppl. 1 (2005): S144–50.

99 **They were a component:** For a recent assessment see Paolo Boffetta and Fredrik Nyberg, "Contribution of Environmental Factors to Cancer Risk," *BMJ: British Medical Journal* 68, no. 1 (December 1, 2003): 71–94; and Richard W. Clapp and Molly M. Jacobs, "Environmental and Occupational Causes of Cancer: New Evidence, 2005–2007," October 2007, Lowell Center for Sustainable Production website.

99 **"there is too much ignorance":** Doll and Peto, "Causes of Cancer," 1251.

99 **cancer mortality among people under sixty-five:** Doll and Peto, "Causes of Cancer," 1256. For details see 1281–85 and tables D1 and D3.

99 **largely true for older Americans:** Doll and Peto, "Causes of Cancer," 1256. See table D2 for the overall rate and D4 for specific cancers. There appeared to be an increase in brain cancer deaths and smaller increases in other non-respiratory cancers, but the authors attributed these mostly to better record keeping.

99 **not because we were getting much better:** Doll and Peto, "Causes of Cancer," 1256.

99 **Two smaller studies:** J. Higginson and C. S. Muir, "Environmental Carci-

nogenesis: Misconceptions and Limitations to Cancer Control," *Journal of the National Cancer Institute* 63, no. 6 (December 1979): 1291–98; and E. L. Wynder and G. B. Gori, "Contribution of the Environment to Cancer Incidence: An Epidemiologic Exercise," *Journal of the National Cancer Institute* 58, no. 4 (April 1977): 825–32.

100 **began challenging the report:** Samuel S. Epstein and Joel B. Swartz, "Fallacies of Lifestyle Cancer Theories," *Nature* 289, no. 5794 (January 15, 1981): 127–30.

100 **When lung cancer rates began rising:** David G. Hoel, Devra L. Davis, et al., "Trends in Cancer Mortality in 15 Industrialized Countries, 1969–1986," *Journal of the National Cancer Institute* 84, no. 5 (March 4, 1992): 313–20.

100 **While epidemiologists kept watch:** Bruce Ames's story is told in Proctor, *Cancer Wars,* 136–52.

100 **the Ames test:** Bruce N. Ames et al., "Carcinogens Are Mutagens: A Simple Test System Combining Liver Homogenates for Activation and Bacteria for Detection," *Proceedings of the National Academy of Sciences* 70, no. 8 (August 1973): 2281–85.

100 **By killing esophagus cells:** Alcohol may also increase cancer risk by breaking down into carcinogenic acetaldehyde and through other mechanisms. For a summary see "Alcohol Use and Cancer," American Cancer Society website, last revised January 27, 2012.

101 **a paper in *Science*:** B. N. Ames, "Dietary Carcinogens and Anticarcinogens," *Science* 221, no. 4617 (September 23, 1983): 1256–64.

101 **In 1997, he reported:** B. N. Ames and L. S. Gold, "Environmental Pollution, Pesticides, and the Prevention of Cancer: Misconceptions," *FASEB Journal: Official Publication of the Federation of American Societies for Experimental Biology* 11, no. 13 (November 1997): 1041–52.

101 **In fact he doubted:** Ames and Gold, "Environmental Pollution, Pesticides, and the Prevention of Cancer."

101 **Half of everything tested:** B. N. Ames and L. S. Gold, "Chemical Carcinogenesis: Too Many Rodent Carcinogens," *Proceedings of the National Academy of Sciences* 87, no. 19 (October 1990): 7772–76.

102 **an experiment costing tens of millions of dollars:** *Cancer and the Environment* (Washington, DC: U.S. Department of Health and Human Services, March 2003), 25.

102 **mitogenesis increases mutagenesis:** Ames and Gold, "Chemical Carcinogenesis."

102 **Toxicologists defended the tests:** P. J. Infante, "Prevention Versus Chemophobia: A Defence of Rodent Carcinogenicity Tests," *Lancet* 337, no. 8740 (March 1991): 538–40; P. F. Infante, "Use of Rodent Carcinogenicity Test

Results for Determining Potential Cancer Risk to Humans," *Environmental Health Perspectives* 101, suppl. 5 (December 1993): 143–48; and I. Bernard Weinstein, "Cell Proliferation: Concluding Remarks," *Environmental Health Perspectives* 101, suppl. 5 (December 1993): 159–61.

102 **diverting attention from a genuine problem:** See, for example, Clapp and Jacobs, "Environmental and Occupational Causes of Cancer"; Devra Lee Davis and Joel Schwartz, "Trends in Cancer Mortality: U.S. White Males and Females, 1968–83," *Lancet* 331 (March 1988): 633–36; and Devra Davis, *The Secret History of the War on Cancer* (New York: Basic Books, 2007).

102 **recent report by a White House advisory group:** *Reducing Environmental Cancer Risk: What We Can Do Now,* 2008–2009 Annual Report (Washington, DC: National Cancer Institute, April 2010).

102 **The alternative would be to administer:** *Reducing Environmental Cancer Risk,* 11.

103 **"has been grossly underestimated":** Reducing Environmental Cancer Risk, introductory letter, unpaginated.

103 **many scientists criticized the report:** David C. Holzman, "President's Cancer Panel Stirs Up Environmental Health Community," *Journal of the National Cancer Institute* 102, no. 15 (August 4, 2010): 1106–13.

103 **The National Academy of Sciences has described:** "Toxicity Testing in the 21st Century: A Vision and a Strategy (Washington, DC: National Academies Press, 2007). These ideas are beginning to be embodied in the Environmental Protection Agency's Computational Toxicology Research program.

103 **death rates from cancer did rise gradually:** Ahmedin Jemal et al., "Annual Report to the Nation on the Status of Cancer, 1975–2009, Featuring the Burden and Trends in Human Papillomavirus (HPV)–Associated Cancers and HPV Vaccination Coverage Levels," *Journal of the National Cancer Institute* (January 7, 2013). See table 2. A summary with a link to the full report is available on the National Cancer Institute's SEER (Surveillance, Epidemiology, and End Results) website.

103 **began decreasing modestly:** Adjusted for age, overall cancer mortality was about 199 people per 100,000 in 1975. Ten years later it was 211. By 2009, the latest year for which statistics have been tabulated, it had dropped to 173. See N. Howlader et al., eds., "SEER Cancer Statistics Review," 1975–2009 (Vintage 2009 Populations), National Cancer Institute, Bethesda, MD, based on November 2011 SEER data submission, posted to the SEER website, 2012. The mortality details are in table 2.6 and incidence details in table 2.5.

103 **Incidence rates tell a similar story:** The 2012 report, cited earlier, doesn't break down the rates so finely. I used table 1 of an earlier annual report: Brenda K. Edwards et al., "Annual Report to the Nation on the Status of Cancer,

1975–2006, Featuring Colorectal Cancer Trends and Impact of Interventions," *Cancer* 116, no. 3 (2010): 544–73.

104 **lack of exercise and excess weight are far more to blame:** The American Association for Cancer Research progress report for 2012 attributes 33 percent of cancer to tobacco, 20 percent to excess weight and obesity, 5 percent to lack of exercise, and just 5 percent to diet (figure 9, page 9). The report is on the AACR's Cancer Progress website. The source of the AACR's numbers is Graham A. Colditz, Kathleen Y. Wolin, and Sarah Gehlert. "Applying What We Know to Accelerate Cancer Prevention," *Science Translational Medicine* 4, no. 127 (March 28, 2012): 127rv4.

104 **A twenty-five-year retrospective:** Graham A. Colditz, Thomas A. Sellers, and Edward Trapido, "Epidemiology—Identifying the Causes and Preventability of Cancer?" *Nature Reviews Cancer* 6, no. 1 (January 2006): 75–83.

104 **found comparable numbers:** "Attributable Causes of Cancer in France in the year 2000," International Agency for Research on Cancer website.

104 **neighborhood cancer clusters:** K. J. Rothman, "A Sobering Start for the Cluster Busters' Conference," *American Journal of Epidemiology* 132, no. 1 suppl. (July 1990): S6–13; and Raymond Richard Neutra, "Counterpoint from a Cluster Buster," *American Journal of Epidemiology* 132, no. 1 (July 1, 1990): 1–8. Also see Atul Gawande, "The Cancer Cluster Myth," *New Yorker,* February 8, 1999. For an evocative account of a cancer cluster investigation and the lessons learned, see Dan Fagin, *Toms River: A Story of Science, Folly and Redemption* (New York: Random House, 2013).

104 **even occupational clusters are uncommon:** For an assessment see P. A. Schulte et al., "Investigation of Occupational Cancer Clusters: Theory and Practice," *American Journal of Public Health* 77, no. 1 (January 1987): 52–56.

104 **the same patterns are appearing:** Ahmedin Jemal et al., "Global Cancer Statistics," *CA: A Cancer Journal for Clinicians* 61, no. 2 (2011): 69–90; P. Boyle and B. Levin, eds., *World Cancer Report 2008* (Lyon: International Agency for Research on Cancer, 2008); and World Cancer Research Fund/American Institute for Cancer Research, *Food, Nutrition, Physical Activity, and the Prevention of Cancer: A Global Perspective* (Washington, DC: AICR, 2007). Also see D. Max Parkin et al., "Global Cancer Statistics, 2002," *CA: A Cancer Journal for Clinicians* 55, no. 2 (February 24, 2009): 74–108.

105 **reverse smoking:** J. J. Pindborg et al., "Reverse Smoking in Andhra Pradesh, India: A Study of Palatal Lesions Among 10,169 Villagers," *British Journal of Cancer* 25, no. 1 (March 1971): 10–20.

105 **unpacking the most recent SEER statistics:** Howlader et al., eds., "SEER Cancer Statistics Review." For the highlights see Jemal et al., "Annual Report to the Nation." I also referred to an earlier report, Betsy A. Kohler et al., "An-

nual Report to the Nation on the Status of Cancer, 1975–2007, Featuring Tumors of the Brain and Other Nervous System," *Journal of the National Cancer Institute* 103, no. 9 (May 4, 2011), 1–23.

105 **a decline or leveling off:** Jemal et al., "Annual Report to the Nation."

106 **12.1 cases per 100,000, compared with 62.6:** Howlader et al., eds., "SEER Cancer Statistics Review," table 1.4.

106 **Childhood cancers are among the very rarest:** Howlader et al., eds., "SEER Cancer Statistics Review," table 28.1.

106 **Death rates . . . have fallen to about half:** Howlader et al., eds., "SEER Cancer Statistics Review," table 1.2.

106 **the numbers jump all over the place:** Howlader et al., eds., "SEER Cancer Statistics Review," table 28.2. The figures are for children age fourteen and under. Also see Trevor Butterworth, "Is Childhood Cancer Becoming More Common?" May 28, 2010, Research at Statistical Assessment Service (STATS) website, George Mason University.

106 **Every cancer tells a different story:** For summaries, see Jemal et al. and the American Cancer Society's annual reports, "Cancer Facts & Figures," on the group's website.

107 **What may appear to be a climb:** Martha S. Linet et al., "Cancer Surveillance Series: Recent Trends in Childhood Cancer Incidence and Mortality in the United States," *Journal of the National Cancer Institute* 91, no. 12 (June 16, 1999): 1051–58. Also see "Childhood Cancers," National Cancer Institute website, reviewed January 10, 2008; and Butterworth, "Is Childhood Cancer Becoming More Common?"

107 **prostate, lung, colorectal . . . are all higher:** Howlader et al., eds., "SEER Cancer Statistics Review," tables 1.5 and 1.6.

107 **less cancer than blacks or whites:** Howlader et al., eds., "SEER Cancer Statistics Review," table 2.5.

107 **the incidence of brain cancer:** Howlader et al., eds., "SEER Cancer Statistics Review," table 3.16.

107 **For liver cancer Hawaii tops out:** Howlader et al., eds., "SEER Cancer Statistics Review," table 14.16.

108 **"Nature and nurture affect the probability":** Doll and Peto, "Causes of Cancer," 1204.

CHAPTER 8 "Adriamycin and Posole for Christmas Eve"

109 **Among the chemicals:** National Toxicology Program, *Report on Carcinogens,* 12th ed. (Research Triangle Park, NC: U.S. Department of Health and Human Services, 2011).

109 **First synthesized in 1844:** Michele Peyrone, "Ueber Die Einwirkung Des Ammoniaks Auf Platinchlorür," *Justus Liebigs Annalen Der Chemie* 51, no. 1 (January 27, 2006): 1–29. For a short biography of the discoverer, see George B. Kauffman et al., "Michele Peyrone (1813–1883), Discoverer of Cisplatin," *Platinum Metals Review* 54, no. 4 (Oct 2010): 250–56.

109 **how cells behaved in the presence of electricity:** Barnett Rosenberg, Loretta Van Camp, and Thomas Krigas, "Inhibition of Cell Division in *Escherichia coli* by Electrolysis Products from a Platinum Electrode," *Nature* 205, no. 4972 (February 13, 1965): 698–99. Also see Gregory A. Petsko, "A Christmas Carol," *Genome Biology* 3, no. 1 (2002); and Rebecca A. Alderden, Matthew D. Hall, and Trevor W. Hambley, "The Discovery and Development of Cisplatin," *Journal of Chemical Education* 83 (2006): 728.

110 **"God, you don't often find things like that":** "Interview with Barnett Rosenberg," Sesquicentennial Oral History Project contributor, available online at Michigan State University Archives and Historical Collections, February 2, 2001.

110 **Rosenberg went on to test the molecule's effects:** Barnett Rosenberg, Loretta Vancamp, et al., "Platinum Compounds: A New Class of Potent Antitumour Agents," *Nature* 222, no. 5191 (April 26, 1969): 385–86.

110 **scientists discovered how that works:** See, for example, Huifang Huang et al., "Solution Structure of a Cisplatin-Induced DNA Interstrand Cross-Link," *Science* 270, no. 5243 (December 15, 1995): 1842–45. For a review of chemotherapy drugs and crosslinking, see Andrew J. Deans and Stephen C. West, "DNA Interstrand Crosslink Repair and Cancer," *Nature Reviews Cancer* 11, no. 7 (July 2011): 467–80; and Laurence H. Hurley, "DNA and Its Associated Processes as Targets for Cancer Therapy," *Nature Reviews Cancer* 2, no. 3 (March 2002): 188–200.

111 **the penicillin of cancer:** Stephen Trzaska, "Cisplatin," *Chemical and Engineering News* 83, no. 25 (2005): 3.

111 **sickening side effects:** "Cisplatin," American Cancer Society website, last revised January 14, 2010.

111 **Doxorubicin has its own curious tale:** Klaus Mross, Ulrich Massing, and Felix Kratz, "DNA-Intercalators—The Anthracyclines," in H. M. Pinedo and Carolien Smorenburg, eds., *Drugs Affecting Growth of Tumours* (Basel, Boston: Birkhäuser Verlag, 2006), 19.

111 **push down your white blood cell count:** "Doxorubicin," American Cancer Society website, last revised November 7, 2011.

111 **reports that the risk increases:** Giorgio Minotti et al., "Paclitaxel and Docetaxel Enhance the Metabolism of Doxorubicin to Toxic Species in Human Myocardium," *Clinical Cancer Research* 7, no. 6 (June 1, 2001): 1511–15.

111 **Paclitaxel (or Taxol) was originally isolated:** Frank Stephenson, "A Tale

of Taxol," *Research in Review,* Fall 2002, available online at the Florida State University Office of Research website.

111 **The first chemo agents:** Alfred Gilman and Frederick S. Philips, "The Biological Actions and Therapeutic Applications of the B-Chloroethyl Amines and Sulfides," *Science* 103, no. 2675 (April 5, 1946): 409–36. For more about the story, see Vincent T. DeVita and Edward Chu, "A History of Cancer Chemotherapy," *Cancer Research* 68, no. 21 (November 1, 2008): 8643–53; and Bruce A. Chabner and Thomas G. Roberts. "Chemotherapy and the War on Cancer," *Nature Reviews Cancer* 5, no. 1 (January 1, 2005): 65–72.

112 **covered under the 1993 Chemical Weapons Convention:** *Convention on the Prohibition of the Development, Production, Stockpiling and Use of Chemical Weapons and on Their Destruction* (New York: Organisation for the Prohibition of Chemical Weapons, 2005). Available on the OPCW website.

112 **"UPSC has a propensity":** Alessandro D. Santin et al., "Trastuzumab Treatment in Patients with Advanced or Recurrent Endometrial Carcinoma Overexpressing HER2/neu," *International Journal of Gynecology & Obstetrics* 102 (August 2008): 128–31.

113 **a cancer of older, thinner women:** David M. Boruta II et al., "Management of Women with Uterine Papillary Serous Cancer," *Gynecologic Oncology* 115 (2009): 142–53; Amanda Nickles Fader et al., "An Updated Clinicopathologic Study of Early-stage Uterine Papillary Serous Carcinoma (UPSC)," *Gynecologic Oncology* 115, no. 2 (November 2009): 244–48; C. A. Hamilton et al., "Uterine Papillary Serous and Clear Cell Carcinomas Predict for Poorer Survival Compared to Grade 3 Endometrioid Corpus Cancers," *British Journal of Cancer* 94, no. 5 (March 13, 2006): 642–46; Sunni Hosemann, "Early Uterine Papillary Serous Carcinoma: Treatment Options Tailored to Patient and Disease Characteristics," *OncoLog* 50, nos. 4–5 (April–May 2010): 4-6; and Carsten Gründker, Andreas R. Günthert, and Günter Emons, "Hormonal Heterogeneity of Endometrial Cancer," in Lev M. Berstein and Richard J. Santen, eds., *Innovative Endocrinology of Cancer,* vol. 630 of *Advances in Experimental Medicine and Biology* (New York, NY: Springer, 2008), 166–88.

113 **"There are no risk factors":** Felice Lackman and Peter Craighead, "Therapeutic Dilemmas in the Management of Uterine Papillary Serous Carcinoma," *Current Treatment Options in Oncology* 4, no. 2 (2003): 99–104.

113 **as few as 5 to 10 percent:** Boruta et al., "Management of Women with Uterine Papillary Serous Cancer." Also see Brij M. Sood et al., "Patterns of Failure After the Multimodality Treatment of Uterine Papillary Serous Carcinoma," *International Journal of Radiation Oncology, Biology, Physics* 57, no. 1 (September 1, 2003): 208–16; and Hadassah Goldberg et al., "Outcome After Combined Modality Treatment for Uterine Papillary Serous Carcinoma: A Study

by the Rare Cancer Network," *Gynecologic Oncology* 108, no. 2 (February 2008): 298–305.

113 **I found an essay:** S. J. Gould, "The Median Isn't the Message," *Discover* 6 (June 1985): 40–42.

114 **"All evolutionary biologists know":** Gould, "The Median Isn't the Message."

117 **Should she be getting topotecan?:** Robert W. Holloway, "Treatment Options for Endometrial Cancer: Experience with Topotecan," part 2, *Gynecologic Oncology* 90, no. 3 (September 2003): S28–33.

117 **He attached abstracts from three papers:** Holly H. Gallion et al., "Randomized Phase III Trial of Standard Timed Doxorubicin Plus Cisplatin Versus Circadian Timed Doxorubicin Plus Cisplatin in Stage III and IV or Recurrent Endometrial Carcinoma," *Journal of Clinical Oncology* 21, no. 20 (October 15, 2003): 3808–13; David Scott Miller et al., "A Phase II Trial of Topotecan in Patients with Advanced, Persistent, or Recurrent Endometrial Carcinoma: A Gynecologic Oncology Group Study," *Gynecologic Oncology* 87, no. 3 (December 2002): 247–51; and Scott Wadler et al., "Topotecan Is an Active Agent in the First-line Treatment of Metastatic or Recurrent Endometrial Carcinoma," *Journal of Clinical Oncology* 21, no. 11 (June 1, 2003): 2110–14.

117 **Nancy's oncologist gave us a paper:** Alessandro D. Santin, "HER2/neu Overexpression: Has the Achilles' Heel of Uterine Serous Papillary Carcinoma Been Exposed?" *Gynecologic Oncology* 88, no. 3 (March 2003): 263–65.

117 **receptors that respond to human epidermal growth factors:** The mechanism is a little more convoluted than is often described. See "Targeted Therapies for Breast Cancer Tutorial: Inhibition of HER2," National Cancer Institute website.

117 **It is usually just called HER2:** The awkward name came about after two laboratories discovered the gene independently (in humans and in rats): Alan L. Schechter, Robert A. Weinberg, et al., "The Neu Oncogene: An erb-B-related Gene Encoding a 185,000-Mr Tumour Antigen," *Nature* 312, no. 5994 (December 6, 1984): 513–16; and A. Ullrich et al., "Human Epidermal Growth Factor Receptor cDNA Sequence and Aberrant Expression of the Amplified Gene in A431 Epidermoid Carcinoma Cells," *Nature* 309, no. 5967 (June 31, 1984): 418–25.

118 **A drug called Herceptin:** The story of its development is told in Robert Bazell, *Her-2: The Making of Herceptin, a Revolutionary Treatment for Breast Cancer* (New York: Random House, 1998).

118 **His name had been mentioned in an episode of *The West Wing*:** Lawrence K. Altman, MD, "Very Real Questions for Fictional President," *Doctor's World, New York Times,* October 9, 2001.

CHAPTER 9 Deeper into the Cancer Cell

121 **neatly described by two scientists:** D. Hanahan and R. A. Weinberg, "The Hallmarks of Cancer," *Cell* 100, no. 1 (January 7, 2000): 57–70.

121 **The idea . . . goes back decades:** C. O. Nordling, "A New Theory on the Cancer-inducing Mechanism," *British Journal of Cancer* 7, no. 1 (March 1953): 68–72. Nordling argued that the need for multiple mutations explains why cancer becomes increasingly frequent with age: "If two mutations were required, the frequency of cancer should increase in direct proportion to age. . . . If three mutations were required, a cancer frequency proportional to the second power of age might be expected, with four mutations to the third power of age, and so on." Peter Nowell is often given credit for the first clear description of the idea of cancer as a Darwinian process in "The Clonal Evolution of Tumor Cell Populations," *Science* 194, no. 4260 (October 1, 1976): 23–28. The theory was put on solid footing with landmark experiments on colorectal cancer. See Bert Vogelstein et al., "Genetic Alterations During Colorectal-tumor Development," *New England Journal of Medicine* 319, no. 9 (September 1, 1988): 525–32.

122 **"For decades now":** Hanahan and Weinberg, "The Hallmarks of Cancer" (italics added).

123 **don't necessarily have to occur through mutations:** The seminal paper on epigenetics is Andrew P. Feinberg and Bert Vogelstein, "Hypomethylation Distinguishes Genes of Some Human Cancers from Their Normal Counterparts," *Nature* 301, no. 5895 (January 6, 1983): 89–92. For a historical overview see Andrew P. Feinberg and Benjamin Tycko, "The History of Cancer Epigenetics," *Nature Reviews Cancer* 4, no. 2 (February 2004): 143–53. Epigenetic changes in germ cells—sperm or eggs—might even be passed from parent to child, though the significance of that is uncertain.

124 **found to be mutated in different cancers:** Päivi Peltomäki, "Mutations and Epimutations in the Origin of Cancer," *Experimental Cell Research* 318, no. 4 (February 15, 2012): 299–310.

125 **proposed that cancer actually begins with epigenetic disruptions:** Andrew P. Feinberg, Rolf Ohlsson, and Steven Henikoff, "The Epigenetic Progenitor Origin of Human Cancer," *Nature Reviews Genetics* 7, no. 1 (January 2006): 21–33.

125 **a contentious idea called the cancer stem cell theory:** Piyush B. Gupta, Christine L. Chaffer, and Robert A. Weinberg, "Cancer Stem Cells: Mirage or Reality?" *Nature Medicine* 15, no. 9 (2009): 1010–12; Jerry M. Adams and Andreas Strasser, "Is Tumor Growth Sustained by Rare Cancer Stem Cells or Dominant Clones?" *Cancer Research* 68, no. 11 (June 1, 2008): 4018–21; and

Peter Dirks, "Cancer Stem Cells: Invitation to a Second Round," *Nature* 466, no. 7302 (July 1, 2010): 40–41. The basic idea was suggested as early as 1937 (J. Furth and M. C. Kahn, "The Transmission of Leukaemia of Mice with a Single Cell," *American Journal of Cancer 31* [1937]: 276–82), and cancer stem cells were identified in a blood cancer by Dominique Bonnet and John E. Dick: "Human Acute Myeloid Leukemia Is Organized as a Hierarchy That Originates from a Primitive Hematopoietic Cell," *Nature Medicine* 3, no. 7 (July 1, 1997): 730–37.

125 **the more confusing it seemed:** For a taste of the controversy see John E. Dick, "Looking Ahead in Cancer Stem Cell Research," *Nature Biotechnology* 27, no. 1 (January 2009): 44–46; Elsa Quintana et al., "Efficient Tumour Formation by Single Human Melanoma Cells," *Nature* 456, no. 7222 (December 4, 2008): 593–98; Priscilla N. Kelly et al., "Tumor Growth Need Not Be Driven by Rare Cancer Stem Cells," *Science* 317, no. 5836 (July 20, 2007): 337; Richard P. Hill, "Identifying Cancer Stem Cells in Solid Tumors: Case Not Proven," *Cancer Research* 66, no. 4 (February 15, 2006): 1891–96; and Scott E. Kern and Darryl Shibata, "The Fuzzy Math of Solid Tumor Stem Cells: A Perspective," *Cancer Research* 67, no. 19 (October 1, 2007): 8985–88.

126 **shed their identity and reverted:** One hypothesis is that they would make the transformation through the epithelial-mensenchymal transformation, which is discussed in chapter 6 of this book.

126 **the wave of the future:** Three papers published in August 2012 set off a surge of publicity in favor of the theory along with a skeptical backlash. For a summary, including citations, see Monya Baker, "Cancer Stem Cells Tracked," *Nature* 488, no. 7409 (August 2, 2012): 13–14.

126 **the annual meeting:** American Association for Cancer Research, 102nd Annual Meeting, "Innovation and Collaboration: The Path to Progress," April 2–6, 2011, Orange County Convention Center, Orlando, Florida.

126 **more than 16,000 scientists:** "AACR Hosts Successful 102nd Annual Meeting in Orlando," Previous Annual Meetings, AACR website.

127 **an amazing video flythrough:** High-definition stills and videos in two and three dimensions are available on the Amgen website.

127 **Amgen had been working on an angiogenesis inhibitor:** Beth Y. Karlan et al., "Randomized, Double-Blind, Placebo-Controlled Phase II Study of AMG 386 Combined with Weekly Paclitaxel in Patients with Recurrent Ovarian Cancer," *Journal of Clinical Oncology* 30, no. 4 (February 1, 2012): 362–71.

127 **extended the lives:** The technical term used in the study was "overall survival."

128 **"Judah is going to cure cancer":** Gina Kolata, "A Cautious Awe Greets Drugs That Eradicate Tumors in Mice, *New York Times,* May 3, 1998.

128 **"the most exciting cancer research of my lifetime":** James Watson, "High Hopes on Cancer," *New York Times,* letter to the editor, May 7, 1998.

128 **"remarkable and wonderful":** Kolata, "A Cautious Awe Greets Drugs."

128 **metastasizing more vigorously:** Erika Check Hayden, "Cutting Off Cancer's Supply Lines," *Nature News* 458, no. 7239 (April 8, 2009): 686–87.

128 **add a few months to a patient's life:** Avastin product page, Genentech website.

128 **the Food and Drug Administration . . . revoked approval:** Andrew Pollack, "F.D.A. Revokes Approval of Avastin for Use as Breast Cancer Drug," *New York Times,* November 18, 2011.

129 **standard chemotherapy was accompanied by Herceptin:** Edward H. Romond et al., "Trastuzumab Plus Adjuvant Chemotherapy for Operable HER2-positive Breast Cancer," *New England Journal of Medicine* 353, no. 16 (October 20, 2005): 1673–84. Also see Luca Gianni et al., "Treatment with Trastuzumab for 1 Year After Adjuvant Chemotherapy in Patients with HER2-positive Early Breast Cancer: A 4-year Follow-up of a Randomised Controlled Trial," *Lancet Oncology* 12, no. 3 (March 2011): 236–244.

129 **Genentech could reduce the time to market:** An early end to a clinical trial is not necessarily considered a good thing. See F. Trotta et al., "Stopping a Trial Early in Oncology: For Patients or for Industry?" *Annals of Oncology* 19, no. 7 (July 1, 2008): 1347–53; Margaret McCartney, "Leaping to Conclusions," *BMJ: British Medical Journal* 336, no. 7655 (May 31, 2008): 1213–14; and Victor M. Montori et al., "Randomized Trials Stopped Early for Benefit: A Systematic Review," *JAMA: The Journal of the American Medical Association* 294, no. 17 (November 2, 2005): 2203–9.

129 **a serious risk of congestive heart failure:** A study of twelve thousand women who took Herceptin found that mortality from breast cancer was reduced by one-third but that there was a fivefold increase in the risk of cardiac toxicity. See Lorenzo Moja et al., "Trastuzumab Containing Regimens for Early Breast Cancer," *Cochrane Database of Systematic Reviews* 2012, issue 4, article no. CD006243, published online April 18, 2012.

129 **the "crowning achievement":** Scott A. Stuart, Yosuke Minami, and Jean Y. J. Wang, "The CML Stem Cell: Evolution of the Progenitor," *Cell Cycle* 8, no. 9 (May 1, 2009): 1338–43. For the story of Gleevec, see Terence Monmaney, "A Triumph in the War Against Cancer," *Smithsonian,* May 2011.

129 **by strengthening the body's immunological defenses:** For an overview see Ira Mellman, George Coukos, and Glenn Dranoff, "Cancer Immunotherapy Comes of Age," *Nature* 480, no. 7378 (December 21, 2011): 480–89; Drew M. Pardoll, "The Blockade of Immune Checkpoints in Cancer Immunotherapy," *Nature Reviews Cancer* 12, no. 4 (April 2012): 252–64; and David L. Porter et

al., "Chimeric Antigen Receptor–Modified T Cells in Chronic Lymphoid Leukemia," *New England Journal of Medicine* 365, no. 8 (August 10, 2011): 725–33.

129 **the patient's own immune cells are removed:** In another approach, killed cancer cells are used to vaccinate patients against their own tumors in much the way that inactivated viruses are used to make influenza vaccines.

130 **as precipitously as they have for heart disease?:** Arialdi M. Miniño et al., "Deaths: Final Data for 2008," *National Vital Statistics Reports* 59, no. 10 (December 7, 2011). See figure 6, page 9.

130 **losing the War on Cancer?:** For a measured argument see Sharon Begley, "We Fought Cancer . . . And Cancer Won," *Newsweek,* September 5, 2008.

130 **founder and chairman of the advisory board:** Telome Health Inc. website.

131 **said that she had lost her slide:** The speaker was Lynda Chin and the company is Aveo Oncology. Her husband is Ronald DePinho, who went on to become president of MD Anderson Cancer Center. In 2012 the couple was involved in a dispute over an $18 million grant. The details are reported in Meredith Wadman, "Texas Cancer Institute to Re-review Controversial Grant," *Nature News,* May 31, 2012. Also see Meredith Wadman, "Texas Cancer-centre Head Apologizes for Promoting Stock on Television," *Nature News,* June 1, 2012.

131 **"I really want this stuff to work":** Ervin J. Epstein, plenary talk, American Association for Cancer Research 102nd Annual Meeting, April 3, 2011. He also noted he had been a consultant for Genentech and Novartis and owned some stock in a company called Curis.

131 **pioneering work on viruses and oncogenes:** D. Stehelin, H. E. Varmus, J. M. Bishop, and P. K. Vogt, "DNA Related to the Transforming Gene(s) of Avian Sarcoma Viruses Is Present in Normal Avian DNA," *Nature* 260, no. 5547 (March 11, 1976): 170–73.

131 **some of the most perplexing questions:** Varmus was talking about the Provocative Questions project, which is described on the National Cancer Institute website. Also see Harold Varmus and Ed Harlow, "Science Funding: Provocative Questions in Cancer Research," *Nature* 481, no. 7382 (January 25, 2012): 436–37.

131 **cancer of the heart:** Timothy J. Moynihan, "Heart Cancer: Is There Such a Thing?" Disease and Conditions, Mayo Clinic Health Information website, April 12, 2012.

131 **the results have been surprising:** Michael R. Stratton, Peter J. Campbell, and P. Andrew Futreal, "The Cancer Genome," *Nature* 458, no. 7239 (April 9, 2009): 719–24; and P. Andrew Futreal, Michael R. Stratton, et al., "A Census of Human Cancer Genes," *Nature Reviews Cancer* 4, no. 3 (March 2004): 177–83.

132 **hundreds of mutations may potentially be involved:** For a particularly striking example, see H. Nikki March et al., "Insertional Mutagenesis Identifies Multiple Networks of Cooperating Genes Driving Intestinal Tumorigenesis," *Nature Genetics* 43, no. 12 (2011): 1202–9. Part of the challenge is distinguishing between "driver" and "passenger" mutations. See chapter 12 of this book for details.

132 **the phenomenon of polarization:** For the relationship to cancer see, for example, Minhui Lee and Valeri Vasioukhin, "Cell Polarity and Cancer—Cell and Tissue Polarity as a Non-canonical Tumor Suppressor," *Journal of Cell Science* 121, no. 8 (April 15, 2008): 1141–50.

132 **the many different kinds of cell death:** Melanie M. Hippert, Patrick S. O'Toole, and Andrew Thorburn, "Autophagy in Cancer: Good, Bad, or Both?" *Cancer Research* 66, no. 19 (October 1, 2006): 9349–51; Michael Overholtzer, Joan S. Brugge, et al., "A Nonapoptotic Cell Death Process, Entosis, That Occurs by Cell-in-Cell Invasion," *Cell* 131, no. 5 (November 30, 2007): 966–79; and Peter Vandenabeele et al., "Molecular Mechanisms of Necroptosis: An Ordered Cellular Explosion," *Nature Reviews Molecular Cell Biology* 11, no. 10 (October 1, 2010): 700–14.

133 **the Warburg effect:** The metabolic change, involving glycolysis, was described by Otto Warburg in "On the Origin of Cancer Cells," *Science* 123, no. 3191 (February 24, 1956): 309–14. When carried out in the presence of oxygen, the process is called aerobic glycolysis. The result is increased consumption of glucose, which is why cancer cells light up in PET scans.

133 **take in more of the raw material:** Matthew G. Vander Heiden, Lewis C. Cantley, and Craig B. Thompson, "Understanding the Warburg Effect: The Metabolic Requirements of Cell Proliferation," *Science* 324, no. 5930 (May 22, 2009): 1029–33.

133 **The slow burn of chronic inflammation:** For a good overview see Gary Stix, "Is Chronic Inflammation the Key to Unlocking the Mysteries of Cancer?" *Scientific American,* July 2007, updated online November 9, 2008. More references are in my notes for chapter 10.

133 **molecules called sirtuins:** For a review see Finkel Toren, Chu-Xia Deng, and Raul Mostoslavsky, "Recent Progress in the Biology and Physiology of Sirtuins," *Nature* 460, no. 7255 (July 30, 2009): 587–91.

133 **the genes residing in the microbes:** Steven R. Gill et al., "Metagenomic Analysis of the Human Distal Gut Microbiome," *Science* 312, no. 5778 (June 2, 2006): 1355–59.

134 **a Human Microbiome Project:** Peter J. Turnbaugh et al., "The Human Microbiome Project," *Nature* 449, no. 7164 (October 18, 2007): 804–10.

134 **"'omics":** Joshua Lederberg christened the microbiome, and in a short essay,

"Ome Sweet 'Omics," he commented on the naming phenomenon: *The Scientist* 15, no. 7 (April 2, 2001): 8.

134 **separate the ridiculome from the relevantome:** I thought I had invented these words, but an Internet search turns them up in a PowerPoint presentation: Andrea Califano, Brian Athey, and Russ Altman, "Creating a DBP Community to Enhance the NCBC Biomedical Impact, A National Center for Biomedical Computing Work Group Report," July 18, 2006, National Alliance for Medical Image Computing website.

134 **Horace Freeland Judson's magnificent book:** *The Eighth Day of Creation: Makers of the Revolution in Biology,* expanded ed. (Cold Spring Harbor, NY: Cold Spring Harbor Laboratory Press, 1996).

135 **microRNAs:** Rosalind C. Lee, Rhonda L. Feinbaum, and Victor Ambros, "The *C. elegans* Heterochronic Gene lin-4 Encodes Small RNAs with Antisense Complementarity to lin-14," *Cell* 75 (December 1993): 843–54.

136 **the importance . . . has been overblown:** Harm van Bakel et al., "Most 'Dark Matter' Transcripts Are Associated with Known Genes," *PLOS Biology* 8, no. 5 (May 18, 2010): e1000371; and Richard Robinson, "Dark Matter Transcripts: Sound and Fury, Signifying Nothing?" *PLOS Biology* 8 (May 18, 2010): e1000370.

136 **a sweeping new theory:** Leonardo Salmena, Pier Paolo Pandolfi, et al., "A ceRNA Hypothesis: The Rosetta Stone of a Hidden RNA Language?" *Cell* 146, no. 3 (August 5, 2011): 353–58. The speaker was the lead author, Pier Paolo Pandolfi.

136 **Junk that is not junk:** Even more of the noncoding DNA appears to have found a purpose with the ENCODE project, whose results were announced with an extravagant multimedia website by the journal *Nature*. For an old-fashioned overview of the results see Consortium, The ENCODE Project, "An Integrated Encyclopedia of DNA Elements in the Human Genome," *Nature* 489, no. 7414 (September 6, 2012): 57–74. Upon publication, a backlash ensued from scientists who thought the results, though important, were hyped. See John Timmer, "Most of What You Read Was Wrong: How Press Releases Rewrote Scientific History," in the online publication *Ars Technica,* September 10, 2012.

137 **had published a follow-up:** Douglas Hanahan and Robert A Weinberg, "Hallmarks of Cancer: The Next Generation," *Cell* 144, no. 5 (March 4, 2011): 646–74. The ten-year anniversary of the original "Hallmarks" paper was taken as occasion for a critique: Yuri Lazebnik, "What Are the Hallmarks of Cancer?" *Nature Reviews Cancer* 10, no. 4 (April 1, 2010): 232–33.

CHAPTER 10  The Metabolic Mess

139  **Alexander Fleming discovered penicillin:** A. Fleming, "On the Antibacterial Action of Cultures of a Penicillium, with Special Reference to Their Use in the Isolation of B. Influenzae," *British Journal of Experimental Pathology* 10 (1929): 226–35. The article was republished in *Bulletin of the World Health Organization* 79, no. 8 (2001): 780–90. He described the discovery in his Nobel Prize lecture, December 11, 1945: Alexander Fleming, "Penicillin," in *Nobel Lectures, Physiology or Medicine 1942–1962* (Amsterdam: Elsevier Publishing Company, 1964), which is available on the Nobel Prize website. Unknown to Fleming, scientists before him had also noticed penicillin's effects (see Horace Freeland Judson, *The Search for Solutions* [London: Hutchinson, 1980], 73–75), and historians have cast doubt on the details of the canonical account: Douglas Allchin, "Penicillin and Chance," Sociology, History and Philosophy in Science Teaching Resource Center website, University of Minnesota.

140  **Boys thin from malnutrition:** H. A. Waldron, "A Brief History of Scrotal Cancer," *British Journal of Industrial Medicine* 40, no. 4 (November 1983): 390–401.

140  **"The fate of these people":** Percivall Pott, *The Chirurgical Works of Percival Pott, F.R.S. and Surgeon to St. Bartholomew's Hospital* (London: Printed for T. Lowndes, J. Johnson, G. Robinson, T. Cadell, T. Evans, W. Fox, J. Bew, and S. Hayes, 1783). The book was originally published in 1775, and the quote is from page 178 of a later, expanded edition (London: J. Johnson, 1808).

140  **"I have many times made the experiment":** Potts, *The Chirurgical Works,* 179.

140  **Chimney sweeps on the European continent:** Waldron, "A Brief History."

141  **unknown in Edinburgh:** Robert M. Green, MD, "Cancer of the Scrotum," *Boston Medical and Surgical Journal* 163, no. 2 (November 17, 1910): 755–59.

141  **"from that of a grain of rice":** K. Yamagiwa and K. Ichikawa, "Experimental Study of the Pathogenesis of Carcinoma," *Journal of Cancer Research* 3 (1918): 1–29. Republished along with a short biography of Katsusaburo Yamagiwa in *CA: A Cancer Journal for Clinicians* 27, no. 3 (May/June 1977): 172–81.

141  **"Every city in Italy":** Bernardino Ramazzini, *Diseases of Workers,* translated from the Latin text *De morbis artificum* of 1713 by Wilmer Cave Wright, with an introduction by George Rosen (Chicago: University of Chicago Press, 1940), 191. This edition includes the Latin text on facing pages. Ramazzini wrote about the nuns in a section called "Wet-Nurses," 189–93. Also see J. S. Felton, "The Heritage of Bernardino Ramazzini," *Occupational Medicine* 47, no. 3 (April 1, 1997): 167–79.

142  **nursing of children:** World Cancer Research Fund/American Institute for

Cancer Research, *Food, Nutrition, Physical Activity, and the Prevention of Cancer: A Global Perspective* (Washington, DC: AICR, 2007), 239–42.

142 **Domenico Rigoni-Stern, observed:** I. D. Rotkin, "A Comparison Review of Key Epidemiological Studies in Cervical Cancer Related to Current Searches for Transmissible Agents," *Cancer Research* 33, no. 6 (June 1, 1973): 1353–67; and Joseph Scotto and John C. Bailar, "Rigoni-Stern and Medical Statistics: A Nineteenth-Century Approach to Cancer Research," *Journal of the History of Medicine and Allied Sciences* 24, no. 1 (1969): 65–75.

143 **"The dogma was that cancer":** All quotations from Riboli are from an interview with the author in London, May 12, 2011.

143 **came from laboratory experiments:** Some pioneering research was done in the 1940s by Albert Tannenbaum. See "The Initiation and Growth of Tumors. Introduction. I. Effects of Underfeeding," *American Journal of Cancer* 38 (1940): 335–50. For some later work see D. Kritchevsky et al., "Calories, Fat and Cancer," *Lipids* 21, no. 4 (April 1986): 272–74; D. Kritchevsky, M. M. Weber, and D. M. Klurfeld, "Dietary Fat Versus Caloric Content in Initiation and Promotion of Mammary Tumorigenesis in Rats," *Cancer Research* 44, no. 8 (August 1984): 3174–77; G. A. Boissonneault, C. E. Elson, and M. W. Pariza, "Net Energy Effects of Dietary Fat on Chemically Induced Mammary Carcinogenesis in F344 Rats," *Journal of the National Cancer Institute* 76, no. 2 (February 1986): 335–38; and M. W. Pariza, "Fat, Calories, and Mammary Carcinogenesis: Net Energy Effects," *American Journal of Clinical Nutrition* 45, no. 1 (January 1, 1987): 261–63.

143 **feeding them different amounts and varieties:** G. J. Hopkins and K. K. Carroll, "Relationship Between Amount and Type of Dietary Fat in Promotion of Mammary Carcinogenesis," *Journal of the National Cancer Institute* 62, no. 4 (April 1979): 1009–12.

143 **Diets too rich in salt:** For an overview see Xiao-Qin Wang, Paul D. Terry, and Hong Yan, "Review of Salt Consumption and Stomach Cancer Risk: Epidemiological and Biological Evidence," *World Journal of Gastroenterology* 15, no. 18 (May 14, 2009): 2204–13.

143 **nitrosamines, N-nitroso compounds, and other substances:** See, for example, P. Issenberg, "Nitrite, Nitrosamines, and Cancer," *Federation Proceedings* 35, no. 6 (May 1, 1976): 1322–26; and William Lijinsky, "N-Nitroso Compounds in the Diet," *Mutation Research/Genetic Toxicology and Environmental Mutagenesis* 443, nos. 1–2 (July 15, 1999): 129–38.

144 **review some four thousand studies:** World Cancer Research Fund/American Institute for Cancer Research, *Food, Nutrition, Physical Activity, and the Prevention of Cancer,* 585.

144 **"Diets containing substantial amounts":** World Cancer Research Fund/

American Institute for Cancer Research, *Food, Nutrition, Physical Activity, and the Prevention of Cancer*, 538.

144 **"predominantly plant-based diets"**: World Cancer Research Fund/American Institute for Cancer Research, *Food, Nutrition, Physical Activity, and the Prevention of Cancer*, 522.

144 **"especially rich in cancer-protective chemicals"**: Jane Brody, "Eat Your Vegetables! But Choose Wisely," Personal Health, *New York Times*, January 2, 2001.

145 **the disappointing follow-up**: World Cancer Research Fund/American Institute for Cancer Research, *Food, Nutrition, Physical Activity, and the Prevention of Cancer*. Updates are posted on the organization's Diet and Cancer Report website.

145 **"in no case now is the evidence ... judged to be convincing"**: World Cancer Research Fund/American Institute for Cancer Research, *Food, Nutrition, Physical Activity and the Prevention of Cancer*, 75, 114.

145 **the European Prospective Investigation into Cancer and Nutrition**: Details can be found on the EPIC website.

146 **only the slightest evidence**: Paolo Boffetta et al., "Fruit and Vegetable Intake and Overall Cancer Risk," *Journal of the National Cancer Institute* 102, no. 8 (April 21, 2010): 529–37.

146 **or even of specific cancers**: For citations see the response to the Boffetta paper by Christine Bouchardy, Simone Benhamou, and Elisabetta Rapiti, "Re: Fruit and Vegetable Intake and Overall Cancer Risk in the European Prospective Investigation into Cancer and Nutrition," *Journal of the National Cancer Institute* (December 16, 2010); and T. J. Key, "Fruit and Vegetables and Cancer Risk," *British Journal of Cancer* 104, no. 1 (January 4, 2011): 6–11.

146 **a small protective effect**: Anthony B. Miller et al., "Fruits and Vegetables and Lung Cancer," *International Journal of Cancer* 108, no. 2 (January 10, 2004): 269–76; Heiner Boeing et al., "Intake of Fruits and Vegetables and Risk of Cancer of the Upper Aero-digestive Tract," *Cancer Causes & Control* 17, no. 7 (September 2006): 957–69; and F. L. Büchner et al., "Fruits and Vegetables Consumption and the Risk of Histological Subtypes of Lung Cancer in the European Prospective Investigation into Cancer and Nutrition (EPIC)," *Cancer Causes & Control* 21, no. 3 (March 2010): 357–71.

146 **too early to make more than tentative conjectures**: Key, "Fruit and Vegetables and Cancer Risk."

146 **people who smoke and drink excessively**: M. K. Serdula et al., "The Association Between Fruit and Vegetable Intake and Chronic Disease Risk Factors," *Epidemiology* 7, no. 2 (March 1996): 161–65.

146 **possibly played a small part**: F. J. van Duijnhoven et al., "Fruit, Vegetables,

and Colorectal Cancer Risk," *American Journal of Clinical Nutrition* 89, no. 5 (May 2009): 1441–52.

146 **that too remains in dispute:** Key, "Fruit and Vegetables and Cancer Risk."

146 **"overly optimistic":** Walter C. Willett, "Fruits, Vegetables, and Cancer Prevention: Turmoil in the Produce Section," *Journal of the National Cancer Institute* 102, no. 8 (April 21, 2010): 510–11. He is commenting on Boffetta et al., "Fruit and Vegetable Intake."

146 **the ten-year risk of getting colorectal cancer:** See Teresa Norat et al., "Meat, Fish, and Colorectal Cancer Risk," *Journal of the National Cancer Institute* 97, no. 12 (June 15, 2005): 906–16. The study found protective effects of roughly the same magnitude for eating fish. Similar evidence for fiber was reported in Sheila A. Bingham et al., "Dietary Fibre in Food and Protection Against Colorectal Cancer in the European Prospective Investigation into Cancer and Nutrition," *Lancet* 361, no. 9368 (May 3, 2003): 1496–501.

147 **have come to conflicting conclusions:** See, for example, D. D. Alexander and C. A. Cushing, "Red Meat and Colorectal Cancer: A Critical Summary of Prospective Epidemiologic Studies," *Obesity Reviews* 12, no. 5 (May 2011): e472–493; and Doris S. M. Chan et al., "Red and Processed Meat and Colorectal Cancer Incidence: Meta-Analysis of Prospective Studies," *PLOS ONE* 6, no. 6 (June 6, 2011). For earlier work see Scott Gottlieb, "Fibre Does Not Protect Against Colon Cancer," *BMJ: British Medical Journal* 318, no. 7179 (January 30, 1999): 281; and C. S. Fuchs, W. C. Willett, et al., "Dietary Fiber and the Risk of Colorectal Cancer and Adenoma in Women," *New England Journal of Medicine* 340, no. 3 (January 21, 1999): 169–76. The World Cancer Research Fund concludes on its Diet and Cancer Report website that the case for fiber is getting stronger.

147 **older women who had gained 15 to 20 kilograms:** Lahmann et al., "Long-term Weight Change and Breast Cancer."

147 **fatness itself . . . appeared to be the driving force:** See, for example, P. H. Lahmann et al., "Long-term Weight Change and Breast Cancer Risk," *British Journal of Cancer* 93, no. 5 (September 5, 2005): 582–89; and Tobias Pischon et al., "Body Size and Risk of Renal Cell Carcinoma in the European Prospective Investigation into Cancer and Nutrition," *International Journal of Cancer* 118, no. 3 (February 1, 2006): 728–38.

147 **as much of 25 percent of cancer:** Graham A. Colditz, Kathleen Y. Wolin, and Sarah Gehlert, "Applying What We Know to Accelerate Cancer Prevention," *Science Translational Medicine* 4, no. 127 (March 28, 2012): 127rv4.

148 **The reasons are complex:** Other important components include the hormone leptin, which is involved in regulating appetite, sex hormone–binding globulins, aromatase (also known as estrogen synthase), and PI3 kinase. See

Sandra Braun, Keren Bitton-Worms, and Derek LeRoith, "The Link Between the Metabolic Syndrome and Cancer," *International Journal of Biological Sciences* (2011): 1003–15; and Stephanie Cowey and Robert W. Hardy, "The Metabolic Syndrome," *American Journal of Pathology* 169, no. 5 (November 2006): 1505–22. Also involved is the Warburg effect, in which cancer cells shift to an essentially anaerobic metabolism. For an overview see Gary Taubes, "Unraveling the Obesity-Cancer Connection," *Science* 335, no. 6064 (January 6, 2012): 28–32.

148 **age of menarche decreases:** See Sandra Steingraber, "The Falling Age of Puberty in U.S. Girls," August 2007, Breast Cancer Fund website, which includes citations to the research, and Sarah E. Anderson, Gerard E. Dallal, and Aviva Must, "Relative Weight and Race Influence Average Age at Menarche," part 1, *Pediatrics* 111, no. 4 (April 2003): 844–50.

148 **greater body height:** See, for example, Jane Green et al., "Height and Cancer Incidence in the Million Women Study," *Lancet Oncology* 12, no. 8 (August 2011): 785–94.

149 **also affects the immune system:** For a review, see Lisa M. Coussens and Zena Werb, "Inflammation and Cancer," *Nature* 420, no. 6917 (December 19, 2002): 860–67; and Gary Stix, "Is Chronic Inflammation the Key to Unlocking the Mysteries of Cancer?" *Scientific American,* July 2007, updated online November 9, 2008.

149 **Rudolf Virchow suggested:** Coussens and Werb, "Inflammation and Cancer."

149 **aspirin and other anti-inflammatory drugs:** See, for example, Peter M. Rothwell et al., "Effect of Daily Aspirin on Risk of Cancer Metastasis: A Study of Incident Cancers During Randomised Controlled Trials," *The Lancet* 379, no. 9826 (April 2012): 1591–1601; and Peter M. Rothwell et al., "Short-term Effects of Daily Aspirin on Cancer Incidence, Mortality, and Non-vascular Death: Analysis of the Time Course of Risks and Benefits in 51 Randomised Controlled Trials," *The Lancet* 379, no. 9826 (April 2012): 1602–12.

149 **"low grade chronic inflammatory state":** See, for example, World Cancer Research Fund/American Institute for Cancer Research, *Food, Nutrition, Physical Activity,* 39, box 2.4.

149 **"wounds that do not heal":** H. F. Dvorak, "Tumors: Wounds That Do Not Heal; Similarities Between Tumor Stroma Generation and Wound Healing," *New England Journal of Medicine* 315, no. 26 (December 25, 1986): 1650–59. A few researchers have been exploring the possibility that red meat might encourage colon cancer because it contains, among other carcinogens, a molecule that elicits an inflammatory immune response. See Maria Hedlund et al., "Evidence for a Human-specific Mechanism for Diet and Antibody-mediated Inflammation in Carcinoma Progression," *Proceedings of the National Academy*

*of Sciences* 105, no. 48 (December 2, 2008): 18936–41; and Pam Tangvoranun-takul et al., "Human Uptake and Incorporation of an Immunogenic Nonhuman Dietary Sialic Acid," *Proceedings of the National Academy of Sciences* 100, no. 21 (October 14, 2003): 12045–50.

149 **also been tied to metabolic syndrome and diabetes:** Kathryn E. Wellen and Gökhan S. Hotamisligil, "Inflammation, Stress, and Diabetes," *Journal of Clinical Investigation* 115, no. 5 (May 2, 2005): 1111–19.

149 **diabetes recedes:** See, for example, Hutan Ashrafian et al., "Metabolic Surgery and Cancer: Protective Effects of Bariatric Procedures," *Cancer* 117, no. 9 (May 1, 2011): 1788–99.

149 **"shiftwork that involves circadian disruption":** Kurt Straif et al., "Carcinogenicity of Shift-work, Painting, and Fire-fighting," *Lancet Oncology* 8, no. 12 (December 2007): 1065–66. The article provides pointers to the epidemiological and laboratory studies considered by WHO's International Agency for Research on Cancer.

150 **"Now it is between fifty and sixty kilograms":** The United States Department of Agriculture has estimated that Americans consume 150 pounds a year of various sugars, including high fructose corn syrup. See *Agriculture Factbook 2001–2002* (Washington, DC: U.S. Department of Agriculture, March 2003), 20.

150 **who argues that carbs and sugar:** See Taubes's books *Good Calories, Bad Calories: Fats, Carbs, and the Controversial Science of Diet and Health* (New York: Vintage, 2008) and *Why We Get Fat: And What to Do About It* (New York: Knopf, 2010).

150 **reducing your energy intake and therefore your insulin load:** For the effect of fiber on insulin secretion see, for example, J. G. Potter et al., "Effect of Test Meals of Varying Dietary Fiber Content on Plasma Insulin and Glucose Response," *American Journal of Clinical Nutrition* 34, no. 3 (March 1, 1981): 328–34.

150 **people are able to lead more sedentary lives:** Even that, however, is controversial. See Herman Pontzer et al., "Hunter-Gatherer Energetics and Human Obesity," *PLOS ONE* 7, no. 7 (July 25, 2012): e40503.

151 **An official statement from EPIC:** "Key Findings," EPIC Project website.

CHAPTER 11   Gambling with Radiation

153 **a wake of corrosive free radicals:** Hongning Zhou et al., "Consequences of Cytoplasmic Irradiation: Studies from Microbeam," *Journal of Radiation Research* 50, suppl. A (2009): A59–A65.

153 **send signals to neighboring cells:** Hongning Zhou et al., "Induction of a

Bystander Mutagenic Effect of Alpha Particles in Mammalian Cells," *Proceedings of the National Academy of Sciences* 97, no. 5 (February 29, 2000): 2099–104.

154 **13.4 percent, may be radon related:** Office of Radiation and Indoor Air, *EPA Assessment of Risks from Radon in Homes* (Washington, DC: United States Environmental Protection Agency, June 2003), iv, available on the EPA website.

154 **smoking is also a factor:** "Radon and Cancer," National Cancer Institute website.

154 **The EPA's scale:** "Health Risks," EPA website, last updated June 26, 2012.

154 **clusters of two neutrons and two protons:** These are, in fact, helium nuclei. Early on it was noticed that radium emits helium as it decays. See William Ramsay and Frederick Soddy, "Experiments in Radioactivity, and the Production of Helium from Radium," *Proceedings of the Royal Society* 72 (1903): 204–7.

156 **70 percent of their time at home:** *EPA Assessment of Risks,* 7, 44.

156 **The chance of a nonsmoker getting lung cancer:** Rebecca Goldin, "Lung Cancer Rates: What's Your Risk?" March 08, 2006, Research at Statistical Assessment Service (STATS) website, George Mason University.

156 **a laboratory analysis of my eyeglasses:** R. L. Fleischer et al., "Personal Radon Dosimetry from Eyeglass Lenses," *Radiation Protection Dosimetry* 97, no. 3 (November 1, 2001): 251–58.

156 **a method using ordinary household glass:** R. W. Field et al., "Intercomparison of Retrospective Radon Detectors," *Environmental Health Perspectives* 107, no. 11 (November 1999): 905–10; D. J. Steck, R. W. Field, et al., "210Po Implanted in Glass Surfaces by Long Term Exposure to Indoor Radon," *Health Physics* 83, no. 2 (August 2002): 261–71; and Kainan Sun, Daniel J. Steck, and R. William Field, "Field Investigation of the Surface-deposited Radon Progeny as a Possible Predictor of the Airborne Radon Progeny Dose Rate," *Health Physics* 97, no. 2 (August 2009): 132–44.

156 **as long as they have owned the objects:** For a few decades, anyway. The half-life of 210Po, one of the radon products that is measured, is twenty-two years.

157 **houses in Grand Junction:** Leonard A. Cole, *Element of Risk: The Politics of Radon* (New York: Oxford University Press, 1994), 10–12.

157 **a construction engineer named Stanley Watras:** Cole, *Element of Risk,* 12.

157 **A study in Winnipeg:** E. G. Létourneau et al., "Case-Control Study of Residential Radon and Lung Cancer," *American Journal of Epidemiology* 140, no. 4 (1994): 310–22.

157 **compared the average radon levels:** These and other studies are summarized in the Winnipeg paper.

157 **a negative correlation:** B. L. Cohen, "Test of the Linear-No Threshold Theory of Radiation Carcinogenesis for Inhaled Radon Decay Products," *Health Physics* 68, no. 2 (February 1995): 157–74.

157 **the study was flawed:** J. H. Lubin, "On the Discrepancy Between Epidemiologic Studies in Individuals of Lung Cancer and Residential Radon and Cohen's Ecologic Regression," *Health Physics* 75, no. 1 (July 1998): 4–10.

157 **skewed by an inverse connection:** J. S. Puskin, "Smoking as a Confounder in Ecologic Correlations of Cancer Mortality Rates with Average County Radon Levels," *Health Physics* 84, no. 4 (April 2003): 526–32. For a sampling of the debate that erupted see B. J. Smith, R. W. Field, and C. F. Lynch, "Residential 222Rn Exposure and Lung Cancer: Testing the Linear No-threshold Theory with Ecologic Data," *Health Physics* 75, no. 1 (July 1998): 11–17; and B. J. Cohen, "Response to Criticisms of Smith et al.," *Health Physics* 75, no. 1 (July 1998): 23–28, 31–33. Rejoinders and rejoinders to the rejoinders followed.

157 **Perhaps cigarette smoke interfered with the radon monitors:** R. W. Field, e-mail to author, June 7, 2012.

158 **hundreds to thousands of picocuries per liter:** Cole, *Element of Risk*, 28.

158 **lung cancer rates among uranium miners:** The studies are summarized in "EPA's Assessment of Risks from Radon," 8, 11, and in Committee on Health Risks of Exposure to Radon (BEIR VI), National Research Council, *Health Effects of Exposure to Radon: BEIR VI* (Washington, DC: The National Academies Press, 1999), 76–78.

158 **how long or how often they had smoked:** *BEIR VI*, 77, table 3-2.

158 **a committee of the National Research Council:** *BEIR VI*, 18.

158 **The most ambitious study:** R. W. Field et al., "Residential Radon Gas Exposure and Lung Cancer: The Iowa Radon Lung Cancer Study," *American Journal of Epidemiology* 151, no. 11 (June 1, 2000): 1091–102.

159 **about 62 cases per 100,000 men and women:** "SEER Stat Fact Sheets: Lung and Bronchus," Surveillance, Epidemiology, and End Results (SEER) Program website.

159 **Three of the analyses:** S. Darby et al., "Radon in Homes and Risk of Lung Cancer: Collaborative Analysis of Individual Data from 13 European Case-control Studies," *BMJ: British Medical Journal* 330, no. 7485 (January 29, 2005): 223. The results are described in Hajo Zeeb and Ferid Shannoun, eds., *WHO Handbook on Indoor Radon: A Public Health Perspective* (Geneva: World Health Organization, 2009), 12.

159 **consider the matter clinched:** For an overview, see Jonathan M. Samet, "Radiation and Cancer Risk: A Continuing Challenge for Epidemiologists," *Environmental Health* 10, suppl. 1 (April 5, 2011): S4.

159 **small doses of radiation are . . . beneficial:** Alexander M. Vaiserman,

"Radiation Hormesis: Historical Perspective and Implications for Low-Dose Cancer Risk Assessment," *Dose-Response* 8, no. 2 (January 18, 2010): 172–91. Also see Edward J. Calabrese and Linda A. Baldwin, "Toxicology Rethinks Its Central Belief," *Nature* 421, no. 6924 (February 13, 2003): 691–92; L. E. Feinendegen, "Evidence for Beneficial Low Level Radiation Effects and Radiation Hormesis," *British Journal of Radiology* 78, no. 925 (January 1, 2005): 3–7; and Jocelyn Kaiser, "Sipping from a Poisoned Chalice," *Science* 302, no. 5644 (October 17, 2003): 376–79.

159 **A Johns Hopkins researcher recently concluded:** Richard E. Thompson, "Epidemiological Evidence for Possible Radiation Hormesis from Radon Exposure," *Dose-Response* 9, no. 1 (December 14, 2010): 59–75.

159 **low-level x-ray, gamma, and beta radiation:** Bobby R. Scott et al., "Radiation-stimulated Epigenetic Reprogramming of Adaptive-response Genes in the Lung: An Evolutionary Gift for Mounting Adaptive Protection Against Lung Cancer," *Dose-Response* 7, no. 2 (2009): 104–31.

160 **the Chernobyl nuclear power plant:** "Chernobyl's Legacy: Health, Environmental and Socio-economic Impacts," The Chernobyl Forum: 2003–2005, 2nd revised version, 2012. Available on the International Atomic Energy Agency website.

160 **an increase in thyroid cancer:** For a recent follow-up see Alina V. Brenner et al., "I-131 Dose Response for Incident Thyroid Cancers in Ukraine Related to the Chernobyl Accident," *Environmental Health Perspectives* 119, no. 7 (July 2011): 933–39.

160 **the biggest public health problem . . . has been psychological:** "Chernobyl's Legacy," 36.

160 **"a paralyzing fatalism":** Elisabeth Rosenthal, "Experts Find Reduced Effects of Chernobyl," *New York Times,* September 6, 2005.

160 **recently opened the Chernobyl site to tourism:** Peter Walker, "Chernobyl: Now Open to Tourists," *The Guardian,* December 13, 2010.

160 **a mecca for wildlife:** Robert J. Baker and Ronald K. Chesser, "The Chernobyl Nuclear Disaster and Subsequent Creation of a Wildlife Preserve," *Environmental Toxicology and Chemistry* 19, no. 5 (May 1, 2000): 1231–32.

160 **killed at least 150,000 people:** "How Many People Died as a Result of the Atomic Bombings?" Radiation Effects Research Foundation website.

160 **527 excess deaths from solid cancers:** Kotaro Ozasa et al., "Studies of the Mortality of Atomic Bomb Survivors, Report 14, 1950–2003: An Overview of Cancer and Noncancer Diseases," *Radiation Research* 177, no. 3 (March 2012): 229–43.

160 **and 103 from leukemias:** David Richardson et al., "Ionizing Radiation and Leukemia Mortality Among Japanese Atomic Bomb Survivors, 1950–2000,"

*Radiation Research* 172, no. 3 (September 2009): 368–82. Using incidence instead of mortality figures, the Radiation Effects Research Foundation attributed 1,900 cases of cancer to the bombs. See "How Many Cancers in Atomic-bomb Survivors are Attributable to Radiation?" on the foundation's website.

160 **Tsutomu Yamaguchi survived both blasts:** Mark McDonald, "Tsutomu Yamaguchi, Survivor of 2 Atomic Blasts, Dies at 93," *New York Times,* January 7, 2010.

161 **"cancer in a molten, liquid form":** Siddhartha Mukherjee, *The Emperor of All Maladies: A Biography of Cancer* (New York: Scribner, 2010), 16.

161 **reburied with Pierre in the Panthéon:** Nanny Fröman, "Marie and Pierre Curie and the Discovery of Polonium and Radium," December 1, 1996, Nobel Prize website.

161 **worried that her body would be dangerously radioactive:** D. Butler, "X-rays, Not Radium, May Have Killed Curie," *Nature* 377, no. 6545 (September 14, 1995): 96.

161 **kept in a lead box at the Bibliothèque Nationale:** Fröman, "Marie and Pierre Curie."

161 **too ill to travel to Stockholm:** Marie Curie, *Pierre Curie (With the Autobiographical Notes of Marie Curie),* trans. Charlotte Kellogg (New York: Macmillan Co., 1923), 125.

162 **Pierre described an experiment:** "Radioactive Substances, Especially Radium," June 6, 1905, in *Nobel Lectures, Physics 1901–1921* (Amsterdam: Elsevier Publishing Company, 1967). Available on the Nobel Prize website.

163 **A targeted drug called Alpharadin:** Christopher Parker et al., "Overall Survival Benefit of Radium-223 Chloride (Alpharadin) in the Treatment of Patients with Symptomatic Bone Metastases in Castration-Resistant Prostate Cancer," 7th NCRI Cancer Conference, November 2011, Liverpool. Also see Deborah A. Mulford, David A. Scheinberg, and Joseph G. Jurcic, "The Promise of Targeted Alpha-particle Therapy," *Journal of Nuclear Medicine* 46 suppl. 1 (January 2005): 199S–204S.

CHAPTER 12  The Immortal Demon

167 **"Kisses for the Cure":** Anne Landman, "How Breast Cancer Became Big Business," PR Watch website, June 14, 2008.

167 **Stand Up to Cancer telethon:** "The Show," Stand Up to Cancer website. (There has since been a 2012 broadcast.)

169 **a workshop that evening at the Parker House:** "Translational Cancer Re-

search for Basic Scientists Workshop," American Association for Cancer Research, October 17–22, 2010, Boston, MA.

170 **told the story of two cousins:** Amy Harmon, "New Drugs Stir Debate on Rules of Clinical Trials," *New York Times,* September 18, 2010. For more about the trial see Amy Harmon, "Target Cancer," a series of six articles, *New York Times,* February 22, 2010, to January 20, 2011.

170 **a mutation in a gene called *BRAF*:** As a result, the gene produces a distorted version of a protein that is part of a cellular growth pathway. Normally the BRAF protein is actuated only when it interacts with another protein called RAS, but the mutation frees it of this constraint. See "Vemurafenib," New Treatments, Melanoma Foundation website. For a description of the cancer and the vemurafenib trials see Paul B. Chapman et al., "Improved Survival with Vemurafenib in Melanoma with BRAF V600E Mutation," *New England Journal of Medicine* 364, no. 26 (June 30, 2011): 2507–16. For a later study see Jeffrey A. Sosman et al., "Survival in BRAF V600-Mutant Advanced Melanoma Treated with Vemurafenib," *New England Journal of Medicine* 366, no. 8 (2012): 707–14.

171 **Phase III proved so definitive:** Andrew Pollack, "Two New Drugs Show Promise in Slowing Advanced Melanoma," *New York Times,* June 6, 2011.

171 **typically living four months longer:** The median overall survival was 13.2 vs. 9.6 months for dacarbazine. See Paul B. Chapman et al., "Updated Overall Survival (OS) Results for BRIM-3," 2012 ASCO Annual Meeting, *Journal of Clinical Oncology* 30 no. 18, suppl. (June 20, 2012): abstract 8502.

171 **sixty-six in the dacarbazine group:** "Clinical Trial Result Information," protocol number NO25026, January 4, 2011, Roche trials database website.

171 **half of the people . . . were dead:** Chapman, "Updated Overall Survival (OS) Results."

171 **"he was running a race":** Alexander Solzhenitsyn, *Cancer Ward,* trans. Nicholas Bethell and David Burg (New York: Farrar, Straus and Giroux, 1969), 250.

171 **through a fortuitous mutation:** Ramin Nazarian et al., "Melanomas Acquire Resistance to B-RAF(V600E) Inhibition by RTK or N-RAS Upregulation," *Nature* 468, no. 7326 (November 24, 2010): 973–77.

171 **a paradoxical side effect:** Fei Su et al., "RAS Mutations in Cutaneous Squamous-cell Carcinomas in Patients Treated with BRAF Inhibitors," *New England Journal of Medicine* 366, no. 3 (January 19, 2012): 207–15.

171 **experimenting with combinations:** In 2012, *The New England Journal of Medicine* reported encouraging results from a trial involving dabrafenib, a different BRAF inhibitor. It was combined with trametinib, which inhibits

MEK, another enzyme in the same cellular pathway. See Keith T. Flaherty et al., "Combined BRAF and MEK Inhibition in Melanoma with BRAF V600 Mutations," *New England Journal of Medicine* (published online September 29, 2012).

172 **described the jarring effect:** Tom Curran, "Oncology as a Team Sport," Translational Cancer Research Workshop, October 17, 2010.

172 **discovered a gene called reelin:** G. D'Arcangelo, T. Curran, et al., "A Protein Related to Extracellular Matrix Proteins Deleted in the Mouse Mutant Reeler," *Nature* 374, no. 6524 (April 20, 1995): 719–23; and G. G. Miao, T. Curran, et al., "Isolation of an Allele of Reeler by Insertional Mutagenesis," *Proceedings of the National Academy of Sciences* 91, no. 23 (November 8, 1994): 11050–54.

172 **8 cases in 10 million:** Betsy A. Kohler et al., "Annual Report to the Nation on the Status of Cancer, 1975–2007, Featuring Tumors of the Brain and Other Nervous System," Journal of the National Cancer Institute 103, no. 9 (May 4, 2011), 1–23, table 5.

172 **5 cases per 100,000 among children:** Kohler et al., "Annual Report," 12, table 6.

172 **the most common pediatric brain tumor:** Charles M. Rudin et al., "Treatment of Medulloblastoma with Hedgehog Pathway Inhibitor GDC-0449," *New England Journal of Medicine* 361, no. 12 (September 17, 2009): 1173–78.

172 **median age of diagnosis is five:** Rudin, "Treatment of Medulloblastoma."

172 **"a clumsy, staggered walking pattern":** "Medulloblastoma," American Brain Tumor Association website (2006), 6.

172 **the five-year survival rate was as high as 80 percent:** "Medulloblastoma," 17.

172 **"I met one kid, a teenager":** Curran, "Oncology as a Team Sport."

173 **research from other labs:** For an overview see Ken Garber, "Hedgehog Drugs Begin to Show Results," *Journal of the National Cancer Institute* 100, no. 10 (May 21, 2008): 692–97.

173 **the story of the cyclopean lambs:** This is told in chapter 6 of this book.

173 **a meeting on brain genetics and development:** Genetic Basis of Brain Development and Dysfunction, March 18–23, 2000, Sagebrush Inn and Conference Center, Taos, New Mexico. The authority on hedgehog signaling was Andrew McMahon at Harvard University.

173 **Curran went on to show:** Justyna T. Romer, T. Curran, et al., "Suppression of the Shh Pathway Using a Small Molecule Inhibitor," *Cancer Cell* 6, no. 3 (September 2004): 229–40.

173 **inhibited bone development:** Garber, "Hedgehog Drugs Begin to Show Results."

173 **there were signs that the drug . . . was safe:** "Experimental Targeted Ther-

apy Shows Early Promise Against Medulloblastomas," St. Jude Children's Research Hospital website, June 5, 2010.

174 **approved for basal cell carcinoma:** "FDA Approval for Vismodegib," National Cancer Institute website.

174 **"forward looking":** Emmy Wang, senior manager, corporate relations, Genentech, e-mail to author on behalf of Fred de Sauvage, March 2, 2012.

174 **a body called the United States Adopted Names Council:** For a deciphering of generic drug names see "USAN Stem List," American Medical Association website.

174 **described the latest findings:** José Baselga, keynote (untitled), Translational Cancer Research Workshop, Boston, October 17, 2010.

174 **"super Herceptin" or trastuzumab emtansine:** Ion Niculescu-Duvaz, "Trastuzumab Emtansine, an Antibody-drug Conjugate for the Treatment of HER2+ Metastatic Breast Cancer," *Current Opinion in Molecular Therapeutics* 12, no. 3 (June 2010): 350–60.

175 **drug called pertuzumab:** Cormac Sheridan, "Pertuzumab to Bolster Roche/Genentech's Breast Cancer Franchise?" *Nature Biotechnology* 29, no. 10 (October 13, 2011): 856–58.

175 **pertuzumab became Perjeta:** "FDA Approval for Pertuzumab," National Cancer Institute website, June 11, 2012.

175 **patients were outraged:** Robert Weisman, "Limits on Test Drugs Add to Patients' Ordeals," *Boston Globe,* January 5, 2011.

175 **the agency insisted on waiting:** Martin de Sa'Pinto and Katie Reid, "FDA Puts Brakes on Roche, ImmunoGen Cancer Drug," *Reuters,* August 27, 2010.

175 **a rally outside Boston City Hall:** The date was December 6, 2011.

175 **"reduced the risk of cancer worsening":** Media release, June 3, 2012, Roche website. Also see Lisa Hutchinson, "From ASCO—Breast Cancer: EMILIA Trial Offers Hope," *Nature Reviews Clinical Oncology* 9, no. 8 (August 1, 2012): 430. It was approved February 22, 2013, by the FDA and is sold as Kadcyla.

176 **"smoothes and tightens the skin":** Barbara Ehrenreich, "Welcome to Cancerland," *Harper's Magazine,* November 2001. Also see Gayle A. Sulik, *Pink Ribbon Blues: How Breast Cancer Culture Undermines Women's Health,* (New York: Oxford University Press, 2011).

176 **how many lives that saves:** For more on the controversy over breast cancer treatment see Robert A. Aronowitz, *Unnatural History: Breast Cancer and American Society* (Cambridge: Cambridge University Press, 2007); and David Plotkin, "Good News and Bad News About Breast Cancer," *The Atlantic,* June 1998.

176 **A recent epidemiological study of 600,000 women:** P. C. Gøtzsche and M. Nielsen, "Screening for Breast Cancer with Mammography," *The Cochrane*

*Library* 4 (2009). A summary was published on the Cochrane website April 13, 2011.

177  **a disturbing number:** Timothy J. Wilt et al., "Radical Prostatectomy Versus Observation for Localized Prostate Cancer," *New England Journal of Medicine* 367, no. 3 (2012): 203–13. Also see G. Sandblom et al., "Randomised Prostate Cancer Screening Trial: 20 Year Follow-up," *BMJ: British Medical Journal* 342 (March 31, 2011): d1539.

177  **About 70 percent of men in their seventies:** For a review of the autopsy studies see Richard M. Martin, "Commentary: Prostate Cancer Is Omnipresent, but Should We Screen for It?" *International Journal of Epidemiology* 36, no. 2 (April 1, 2007): 278–81.

177  **Urologists offer free tickets:** The examples are from Gary Schwitzer, "Cheerleading, Shibboleths and Uncertainty," a presentation on April 23, 2012, Science Writing in the Age of Denial, University of Wisconsin, Madison, WI. The urinal example was provided to Schwitzer by Ivan Oransky, the executive editor of Reuters Health.

177  **one of "the best and brightest":** Tom Junod, "Franziska Michor Is the Isaac Newton of Biology," *Esquire,* November 20, 2007.

178  **uncovered some of the early clues:** See, for example, J. C. Fisher, "Multiple-Mutation Theory of Carcinogenesis," *Nature* 181 (March 1, 1958): 651–52; P. Armitage and R. Doll, "The Age Distribution of Cancer and a Multi-stage Theory of Carcinogenesis," *British Journal of Cancer* 8 (1954): 1–12; and C. O. Nordling, "A New Theory on the Cancer-inducing Mechanism," *British Journal of Cancer* 7, no. 1 (March 1953): 68–72.

178  **"hitchhiker" or "passenger" mutations:** Ondrej Podlaha, Franziska Michor, et al., "Evolution of the Cancer Genome," *Trends in Genetics* 28, no. 4 (April 1, 2012): 155–63.

178  **the order in which the mutations occur:** Camille Stephan-Otto Attolini, Franziska Michor, et al., "A Mathematical Framework to Determine the Temporal Sequence of Somatic Genetic Events in Cancer," *Proceedings of the National Academy of Sciences* 107, no. 41 (October 12, 2010): 17604–9.

179  **so quickly overcome the obstacles:** Podlaha, Michor, et al., "Evolution of the Cancer Genome."

179  **it sometimes pays for adversaries to cooperate:** The classic paper is R. Axelrod and W. D. Hamilton, "The Evolution of Cooperation," *Science* 211, no. 4489 (March 27, 1981): 1390–96.

179  **has suggested how that might apply:** Robert Axelrod, David E. Axelrod, and Kenneth J Pienta, "Evolution of Cooperation Among Tumor Cells," *Proceedings of the National Academy of Sciences* 103, no. 36 (September 5, 2006): 13474–79.

180 **"Cancer isn't getting smarter":** The Stand Up to Cancer presentation was at the 2011 Annual Meeting of the American Association for Cancer Research, Orlando, FL, April 2–6. The scientist quoted is Angelique Whitehurst.

180 **"an ugly mass, rounded and bulging":** David Quammen, "Contagious Cancer: The Evolution of a Killer," *Harper's Magazine,* April 2008.

180 **scientists have since traced the origin:** Elizabeth P. Murchison et al., "Genome Sequencing and Analysis of the Tasmanian Devil and Its Transmissible Cancer," *Cell* 148, no. 4 (February 17, 2012): 780–91.

180 **"the immortal devil":** Ewen Callaway, "Field Narrows in Hunt for Devil Tumour Genes," *Nature,* News and Comment, published online February 16, 2012.

180 **spread between hamsters by mosquitoes:** W. G. Banfield et al., "Mosquito Transmission of a Reticulum Cell Sarcoma of Hamsters," *Science* 148, no. 3674 (May 28, 1965): 1239–40.

CHAPTER 13 Beware the Echthroi

183 **"See if aspirin cures the headache":** My trip to Sandia Crest and the situation in Santa Fe are described in "On Top of Microwave Mountain," *Slate,* April 21, 2010.

183 **a book about mass hysteria:** Elaine Showalter, *Hystories: Hysterical Epidemics and Modern Media* (New York: Columbia University Press, 1997).

183 **the threshold set by the Federal Communications Commission:** Federal Communications Commission, "Evaluating Compliance with FCC Guidelines for Human Exposure to Radiofrequency Electromagnetic Fields," *OET Bulletin* 65 (August 1997): 67. See part B of table 1, "Limits for Maximum Permissible Exposure (MPE)," for 1,500–100,000 Mhz. (The occupational limit is 5mW/cm2 for six minutes.) For more information see the FCC's "Questions and Answers About Biological Effects and Potential Hazards of Radiofrequency Electromagnetic Fields," *OET Bulletin* 56, 4th ed. (August 1999). Cell phone exposure is also measured in watts per kilogram—the rate of radio frequency energy absorption by the body.

183 **about 100 milliwatts per square centimeter:** "Calculating the Energy from Sunlight over a 12–Hour Period," Math & Science Resources, National Aeronautics and Space Administration website.

184 **contradictory and inconclusive:** For a summary and general information about wireless technologies and health, see Rfcom, a website maintained by the McLaughlin Centre for Population Health Risk Assessment at the University of Ottawa.

184 **A review by the World Health Organization:** "Electromagnetic Fields, Summary of Health Effects," WHO website. Another good source is "Cell Phones and Cancer Risk" on the National Cancer Institute website.

184 **has remained extremely low:** See table 1.4 of the SEER statistics, N. Howlader et al., eds., "SEER Cancer Statistics Review," 1975–2009 (Vintage 2009 Populations), National Cancer Institute, Bethesda, MD, based on November 2011 SEER data submission, posted to the SEER website, 2012.

184 **slightly but steadily decreasing:** N. Howlader et al., eds., "SEER Cancer Statistics Review," table 1.7.

184 **The most ambitious of these efforts:** "The Interphone Study," International Agency for Research on Cancer, World Health Organization, IARC website.

185 **No relationship was found:** "IARC Report to the Union for International Cancer Control (UICC) on the Interphone Study," October 3, 2011, IARC website.

185 **odds of being diagnosed with the cancer:** This was a hard number to come up with. The statistics available online from SEER don't break down brain tumors by type, but the agency made the calculation at my request. (E-mail to author from Rick Borchelt, NCI Media Relations, July 12, 2012.) For a somewhat lower estimate, see table 1 of Judith A. Schwartzbaum et al., "Epidemiology and Molecular Pathology of Glioma," *Nature Clinical Practice Neurology* 2, no. 9 (2006): 494–503. Adding the incidence rates of the different kinds of glioma comes to 0.0049. The article also estimates that 77 percent of primary malignant brain tumors are gliomas. Multiplying SEER's incidence rate for all gliomas, 0.0061, by 0.77 yields a slightly different value, 0.0047.

185 **a later study by the National Cancer Institute:** M. P. Little et al., "Mobile Phone Use and Glioma Risk: Comparison of Epidemiological Study Results with Incidence Trends in the United States," *BMJ: British Medical Journal* 344 (March 8, 2012): e1147.

186 **to add microwaves:** "IARC Classifies Radiofrequency Electromagnetic Fields as Possibly Carcinogenic to Humans," May 31, 2011, IARC website. The IARC classifications are described on the agency's website, last updated March 27, 2012.

186 **COSMOS:** Joachim Schüz et al., "An International Prospective Cohort Study of Mobile Phone Users and Health (Cosmos): Design Considerations and Enrollment," *Cancer Epidemiology* 35, no. 1 (February 2011): 37–43.

186 **suggested to widespread disbelief:** The original study on power lines and cancer is Nancy Wertheimer and Ed Leeper, "Electrical Wiring Configurations and Childhood Cancer," *American Journal of Epidemiology* 109, no. 3 (March 1, 1979): 273–84.

186 **Robert Weinberg once estimated:** Over a lifetime a human body makes on

the order of $10^{16}$ cells. Divide that by the number of seconds in an 80-year life span, or 2.5 x $10^9$, to get 4 x $10^6$. Robert Weinberg, e-mail to author, November 8, 2010. In *Biology of Cancer*, page 43, he gives an order of magnitude estimate of 10 million.

186 **we all would eventually get cancer:** Interview with Robert Weinberg, August 18, 2010, Whitehead Institute, Boston, MA.

187 **cancer is here "on purpose":** Interview with Robert Austin, October 21, 2010, Princeton University. He expanded on this idea at the first workshop organized by the National Cancer Institute's Physical Sciences in Oncology program, "Integrating and Leveraging the Physical Sciences to Open a New Frontier in Oncology," February 26–28, 2008, Arlington, VA.

187 **Maybe . . . the cells in an organism do the same thing:** Guillaume Lambert, Robert H. Austin, et al., "An Analogy Between the Evolution of Drug Resistance in Bacterial Communities and Malignant Tissues," *Nature Reviews Cancer* 11, no. 5 (April 21, 2011): 375–82.

187 **an attempt to break the stalemate:** The program is called Physical Sciences in Oncology. See Franziska Michor et al., "What Does Physics Have to Do with Cancer?" *Nature Reviews Cancer* 11, no. 9 (August 18, 2011): 657–70; and Paul Davies, "Rethinking Cancer," *Physics World* (June 2010): 28–33.

187 **studying the mechanical forces:** Denis Wirtz, Konstantinos Konstantopoulos, and Peter C. Searson, "The Physics of Cancer: The Role of Physical Interactions and Mechanical Forces in Metastasis," *Nature Reviews Cancer* 11, no. 7 (June 24, 2011): 512–22.

187 **a different level of abstraction:** This was the subject of the Third Physical Sciences in Oncology Workshop, "The Coding, Decoding, Transfer, and Translation of Information in Cancer," October 29–31, 2008, Arlington, VA.

187 **cells can be thought of as oscillators:** Donald Coffey, First Physical Sciences in Oncology Workshop, "Integrating and Leveraging the Physical Sciences."

188 **radio frequency waves to kill cancer cells:** Mustafa Raoof and Steven A. Curley, "Non-Invasive Radiofrequency-Induced Targeted Hyperthermia for the Treatment of Hepatocellular Carcinoma," *International Journal of Hepatology* 2011 (2011): 1–6.

188 **an "ancient genetic toolkit":** Paul Davies, "Cancer: The Beat of an Ancient Drum?" *The Guardian*, April 25, 2011. For a fuller description of the hypothesis, see P. C. W. Davies and C. H. Lineweaver, "Cancer Tumors as Metazoa 1.0: Tapping Genes of Ancient Ancestors," *Physical Biology* 8, no. 1 (February 1, 2011): 015001.

189 **detailed computer simulations:** The USC program, like the others, is described on the National Cancer Institute's Physical Sciences in Oncology website.

189   **a Tinkertoy computer:** A. K. Dewdney, "A Tinkertoy Computer That Plays Tic-tac-toe," *Scientific American* 261, no. 4 (October 1989): 120–23.

189   **a giant clock:** Described on the Long Now Foundation website.

189   **told an audience of oncologists:** Coffey, "Integrating and Leveraging the Physical Sciences."

189   **another of his ambitious machines:** Described on the Applied Proteomics website.

189   **concentrating on the proteome:** Interviews with Daniel Hillis, November 26, 2010, and David Agus, November 29, 2010, Los Angeles.

189   **working for years on mapping the proteome:** See, for example, Bonnie S. Watson et al., "Mapping the Proteome of Barrel Medic (Medicago Truncatula)," *Plant Physiology* 131, no. 3 (March 2003): 1104–23.

190   **continually announcing new discoveries:** See, for example, "Comprehensive Molecular Portraits of Human Breast Tumours," published online in *Nature* (September 23, 2012).

191   **"Ten Crazy Ideas About Cancer":** Seminar at Arizona State University, September 8, 2011. A summary and video are on ASU's Center for the Convergence of Physical Science and Cancer Biology website.

191   **mitochondria . . . might once have been bacteria:** L. Margulis, "Archaeal-eubacterial Mergers in the Origin of Eukarya: Phylogenetic Classification of Life," *Proceedings of the National Academy of Sciences* 93, no. 3 (February 6, 1996): 1071–76.

191   **suspected of playing a part in cancer:** Jennifer S. Carew and Peng Huang, "Mitochondrial Defects in Cancer," *Molecular Cancer* 1, no. 1 (December 9, 2002): 9; and G. Kroemer, "Mitochondria in Cancer," *Oncogene* 25, no. 34 (August 7, 2006): 4630–32.

191   **initiate apoptosis, the cellular suicide routine:** Douglas R. Reed and John C. Green, "Mitochondria and Apoptosis," *Science* 281, no. 5381 (August 28, 1998): 1309–12.

192   **Madeleine L'Engle:** *A Wrinkle in Time* (New York: Farrar, Straus and Giroux, 1962) and *A Wind in the Door* (New York: Farrar, Straus and Giroux, 1973).

193   **a protein that is used as a biomarker:** R. C. Bast Jr. et al., "Reactivity of a Monoclonal Antibody with Human Ovarian Carcinoma," *Journal of Clinical Investigation* 68, no. 5 (November 1981): 1331–37.

193   **a blunt-edged tool:** Charlie Schmidt, "CA-125: A Biomarker Put to the Test," *Journal of the National Cancer Institute* 103, no. 17 (September 7, 2011): 1290–91.

194   **another invader from the Russian steppes:** James A. Young, "Tumbleweed," *Scientific American* 264, no. 3 (March 1991): 82–86.

194 **triclopyr:** "Dow AgroSciences Garlon Family of Herbicides," Dow AgroSciences website.

195 **Maxwell's demon:** I have given only a very general description of the thought experiment devised in the nineteenth century by James Clerk Maxwell, which involved sorting hot and cold gas molecules in a closed chamber. For a collection of essays about the demon and the debate it inspired, see Harvey S. Leff and Andrew F. Rex, *Maxwell's Demon: Entropy, Information, Computing* (Princeton, NJ: Princeton University Press, 1990).


EPILOGUE  Joe's Cancer


198 **said that about 52,000 people:** "Head and Neck Cancers," National Cancer Institute website.

199 **"A melanoblastoma is such a swine":** Alexander Solzhenitsyn, *Cancer Ward,* trans. Nicholas Bethell and David Burg (New York: Farrar, Straus and Giroux, 1969), 202.

199 **a concept called field cancerization:** D. P. Slaughter, H. W. Southwick, and W. Smejkal, "Field Cancerization in Oral Stratified Squamous Epithelium: Clinical Implications of Multicentric Origin," *Cancer* 6, no. 5 (September 1953): 963–68.

200 **"a ticking time bomb":** Boudewijn J. M. Braakhuis et al., "A Genetic Explanation of Slaughter's Concept of Field Cancerization Evidence and Clinical Implications," *Cancer Research* 63, no. 8 (April 15, 2003): 1727–30. For other references on field cancerization see Gabriel D. Dakubo et al., "Clinical Implications and Utility of Field Cancerization," *Cancer Cell International* 7 (2007): 2; and M. G. van Oijen and P. J. Slootweg, "Oral Field Cancerization: Carcinogen-induced Independent Events or Micrometastatic Deposits?" *Cancer Epidemiology, Biomarkers & Prevention* 9, no. 3 (March 2000): 249–56.

202 **William Crookes, the inventor:** W. Crookes, "The Emanations of Radium," *Proceedings of the Royal Society of London* 71 (January 1, 1902): 405–8.

202 **unveiled it at a gala:** Paul W. Frame, "William Crookes and the Turbulent Luminous Sea," Oak Ridge Associated Universities website. The piece originally appeared in the *Health Physics Society Newsletter.*

202 **spinthariscopes with the same engraving:** In Robert Bud and Deborah Jean Warner, eds., *Instruments of Science: An Historical Encyclopedia* (New York: Garland, 1998), 572–73, Helge Kragh writes that the Crookes spinthariscope was produced in the summer of 1903 by several different instrument makers.

202 **"a turbulent, luminous sea":** W. Crookes, "Certain Properties of the Emanations of Radium," *Chemical News* 87, no. 241 (1903).

# Index

A NOTE ABOUT THE AUTHOR

George Johnson has written about science for *The New York Times*, *National Geographic Magazine*, *Slate*, *Scientific American*, *Wired*, *The Atlantic*, and other publications. His nine books, which are being translated into fifteen languages, include *The Ten Most Beautiful Experiments* and *Fire in the Mind: Science, Faith, and the Search for Order*. He has twice been a finalist for the Royal Society's science book prize. A winner of the AAAS Science Journalism Award, he is codirector of the Santa Fe Science Writing Workshop and a former Alicia Patterson fellow. He lives in Santa Fe, New Mexico, and can be found on the Web at talaya.net.

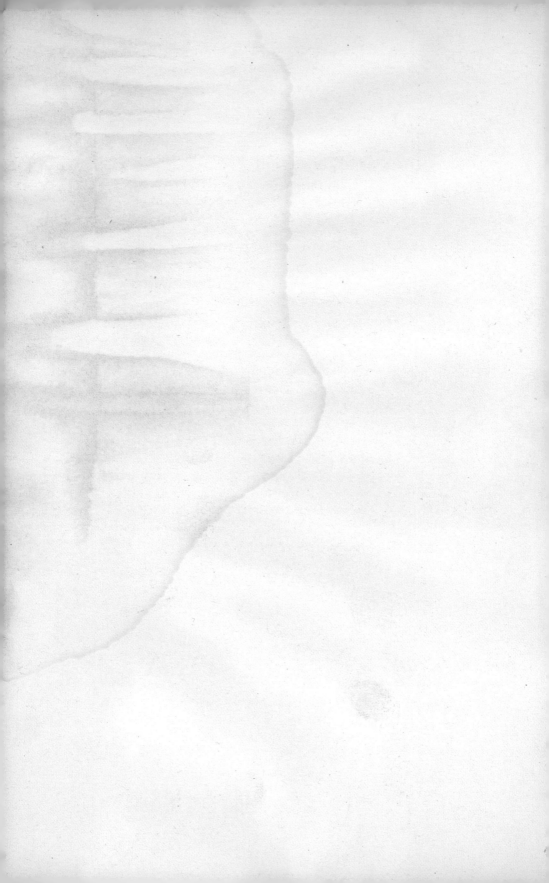